T0270569

DAYDREAMING IN THE SOLAR SYSTEM

# DAYDREAMING IN THE SOLAR SYSTEM: SURFING SATURN'S RINGS, GOLFING ON THE MOON, AND OTHER ADVENTURES IN SPACE EXPLORATION

JOHN E. MOORES AND JESSE ROGERSON
ILLUSTRATED BY MICHELLE D. PARSONS
FOREWORD BY ROBERT J. SAWYER

The MIT Press
Cambridge, Massachusetts
London, England

The MIT Press would like to thank the anonymous peer reviewers who provided comments on drafts of this book. The generous work of academic experts is essential for establishing the authority and quality of our publications. We acknowledge with gratitude the contributions of these otherwise uncredited readers.

This book was set in Bembo Book MT Pro by New Best-set Typesetters Ltd. Printed and bound in the United States of America.

Library of Congress Cataloging-in-Publication Data

Names: Moores, John E., author. | Rogerson, Jesse, 1985– author. | Parsons, Michelle (Michelle D.), illustrator. | Sawyer, Robert J., writer of foreword.
Title: Daydreaming in the solar system : surfing Saturn's rings, golfing on the moon, and other adventures in space exploration / John E. Moores and Jesse Rogerson ; illustrated by Michelle Parsons ; foreword by Robert J. Sawyer.
Description: Cambridge, Massachusetts : The MIT Press, [2024] | Includes bibliographical references and index.
Identifiers: LCCN 2024017190 (print) | LCCN 2024017191 (ebook) | ISBN 9780262049290 (hardcover) | ISBN 9780262380102 (epub) | ISBN 9780262380119 (pdf)
Subjects: LCSH: Outer space—Exploration—Popular works. | Solar system—Popular works.
Classification: LCC QB500.262 .M667 2024 (print) | LCC QB500.262 (ebook) | DDC 919.904—dc23/eng/20240509
LC record available at https://lccn.loc.gov/2024017190
LC ebook record available at https://lccn.loc.gov/2024017191

10   9   8   7   6   5   4   3   2   1

For Elaine and Charlie.
Thank you for being a part of our lives and inspiring us to daydream.

# CONTENTS

FOREWORD

I know the exact moment I decided I wanted to be a science-fiction writer. It happened in 1972, when I was twelve years old. My big brother Peter had given me a beat-up paperback copy of a now long-forgotten young-adult SF novel called *Trouble on Titan* by Alan E. Nourse, first published in 1954.

Those two years are significant: 1972, because that was the last time any human walked on any world other than Earth, and 1954 (seventy years ago now), because back when Nourse wrote that novel, set on the moon of Saturn named in its title, not only had no humans gone into space yet but even the first satellite, Sputnik 1, was still three years in the future.

And so, quite appropriately, Nourse preceded his novel with a 1,000-word introduction (about the same length as the very one you're now reading) entitled "I've Never Been There," explaining that SF is a joy to write in part because, as he said, "There *are* limitations in science fiction which the readers demand, and which the writers must obey."

Nourse then spent the next 250 words giving a recap of what science actually knew all those years ago about Titan, and then he talked about how using that information allowed him to craft a believable, realistic vision of a place no one had yet set foot on. I knew as soon as I finished reading his introduction that I wanted to be a science-fiction writer, too, and that Nourse's brand, rooted in plausibility (what we call "hard SF" in the trade), was the kind I wanted to write.

So, imagine my delight when the manuscript for *Daydreaming in the Solar System* popped into my inbox. John E. Moores and Jesse Rogerson have reversed Nourse's model—giving us fiction first followed by a discussion of the facts. But not being content with just one moon, they instead take us on a fascinating journey to the highlights of the entire solar system (including, of course, Titan). And they've got one up on Nourse: although, like him, they've never been to any of the exotic locales they write about, our space probes have landed on, or at least flown by, most of them, and the data they've beamed back to Earth richly informs the narrative you're about to read.

Until we actually get to these places, the descriptions Moores and Rogerson give—along with this book's charming watercolors by Michelle Parsons—are (to quote the old Bell ads touting long-distance calling) "the next best thing to being there." In these pages, you'll visit the far side of our own Moon (the only place in the solar system that can't ever be seen from Earth), the giant rift valley on Mars that puts the Grand Canyon to shame, the surface of Venus, the rings of Saturn, and more.

Of course, someday humans really will visit these exotic destinations. Although interstellar travel may forever be beyond our grasp (curse you, Albert Einstein!), even the farthest reaches of the solar system will continue to be explored by our space probes, and eventually people will walk (or surf, or skydive, or rappel—all of which happen in this book) on or above the various worlds that make up our sun's family. Even if we never get warp drive, there are indeed plenty of strange new worlds in our own backyard.

Obviously, this is a book aimed at adults—the publisher, after all, is the MIT Press. But, nonetheless, the present volume reminded me of another book I fondly remember from even earlier in childhood: *You Will Go to the Moon* by Mae and Ira Freeman (like *Daydreaming in the Solar System*, also charmingly illustrated, in that case by Robert Patterson). Published in 1959 (although I read it six years later when I was in kindergarten), *You Will Go to the Moon* adopted the same conceit our authors use here: second-person narration. It wasn't just that *someone* would go to the Moon, or to the places explored in this book; rather, it's that *you* will.

Granted, although William Shatner (a Canadian, like the authors of this book and myself) *did* make it to space, I don't expect I will (although, Jeff Bezos or Elon Musk, I'll gladly take your call!). But after reading the scenarios presented here, I now feel as if I have indeed been there.

It used to be that lots of science fiction reveled in the wonders of our solar system. Between 1952 and 1958 Isaac Asimov (writing under the pen name Paul French) published his six Lucky Starr novels, giving us his own fictional tour of the solar system, rooted, as the scenarios in the present volume are, in the best science known at the time. Asimov took us (in order of publication, not in order outward from the sun) to Mars, the asteroids, Venus, Mercury, Jupiter, and Saturn.

Likewise, Robert A. Heinlein gave us *Rocket Ship Galileo* (about a voyage to Earth's moon, published in 1947), *Farmer in the Sky* (1950, about agriculture on Jupiter's moon Ganymede), and *Podkayne of Mars* (1963).

But, these days, rigorously researched science fiction confined to our solar system is rare. Sure, Kim Stanley Robinson wrote a magisterial trilogy about the terraforming of Mars (*Red Mars, Green Mars,* and *Blue Mars*)—but that was thirty years ago. Most of what passes as science fiction these days is just space opera, with people zipping around the galaxy (I blame you, George Lucas!).

Still, as of this writing, the much-lauded TV series *For All Mankind* (ongoing since 2019), with its alternate version of the crewed space program, now has people on a realistic Mars, and the Venus Ascendant novels by Derek Künsken (another Canadian), starting with *The House of Styx* (2021), very convincingly portray life in the clouds of Venus. But if you want what Carl Sagan famously dubbed "the grand tour"—all the highlights of our solar system—*Daydreaming in the Solar System* is the book to read. And I suggest you go ahead and do just that: it's time to turn the page, and, lo and behold, you *will* go to a moon, even if the moon we start with is Jupiter's volcanic satellite Io. Enjoy your trip!

Robert J. Sawyer

Crouched on the surface of Io, the most tightly held of Jupiter's major moons, you feel the rumble long before you see the cause. The ground shakes and the deep vibration travels up your suit's rigid exoskeleton. The suit transfers its energy into your body: first into your legs, then into your gut before moving into your chest. You can feel the power of the volcano deep within you, like some geological rock concert bass line. What a sensation!

Without warning, the dust on your faceplate starts to scatter a flickering glow toward your eyes. You look up and are not disappointed by the accompanying pyrotechnics. The volcano Pele is erupting. The lava lake sputters, sending molten sulfurous material skyward in a fire fountain. In the low gravity, the glowing blobs reach surprising heights and hang suspended above the lake for an unearthly long time.

You hadn't expected this eruption to happen so soon, but you're not surprised. Io is the most volcanically active object in the solar system, and Pele is one of the most active of its volcanoes. Even so, you should be okay. You are protected by your suit. The outer layer is made of golden metal polished to a mirror shine so it can reflect any heat Pele sends your way.

It also doesn't hurt that eruptions are just different out here. With the combination of no air and little gravity, materials emitted by the volcano spread out like the skin of an enormous umbrella. Some of those cinders will rise as high as 300 km above the surface of Io before fanning out along a ballistic trajectory, falling back to Io's surface up to 600 km away. The impossibly massive scale of these fountain-like eruptions even fooled the first spacecraft to visit Jupiter into thinking that more moons were hiding behind Io.

This is one case in which it pays to be closer to the volcano. Most materials will fall in a ring much farther from the volcanic vent than where you are currently standing. Few objects will return to the surface here, deep under the protection of Pele's umbrella. But few doesn't mean none. There's a reason that volcanologist is the most dangerous kind of scientist to be. It's not inconceivable that a house-sized block of glowing rock will come flying over the horizon toward you. But that also keeps things exciting. Nothing entirely safe is worth the rocket flight. And after all, how many human beings have seen something like this from the surface of another world?

You look up and you can already see a change in the sky. Slowly the stars are being blotted out by sunlight and Jupiter-shine scattered by the dust as ash from the plume fills the sky above. Last to be erased will be the disk of Jupiter itself, a massive orb overhead looming with a diameter from this perspective equivalent to nearly thirty-eight full Earth Moons stacked on top of one another. Before you know it, the curtain of materials flowing overhead is complete and the ground below is cast into a wan, yellowish light.

However, for you, the sheer alien-ness of the hellish scene is exactly the point. You want to see something you couldn't find on Earth. A tapestry of natural landscape so extreme that it exists only out here among the planets. Where most people would be running away, looking for escape, you stand still, reveling in the sublime experience. You use this time to gather your courage and your wits. Turning toward the fire fountain at the center of the volcano, you walk forward in awe and amazement. With each measured step, you are eager for the discoveries that await over the next ridgeline . . .

★★

For many of us who make our living in science and engineering, we can often point to a memory where the passion for our work was kindled. Sometimes that moment can come from a real-world experience. Perhaps we come across a particular human object that fascinates in its clockwork complexity and seemingly limitless reconfiguration. Or, instead, an ocean of possibilities rises up before us on the seashore, while staring into a creature's eyes, on the top of a mountain, or beneath a canopy of stars or leaves.

But, to paraphrase the noted neurologist, naturalist, and writer Oliver Sacks, the human mind sees not only with the eyes but also with the imagination. That means that the kind of inspiration that fires up our minds can also be found in books, music, and film. The story of the future astronaut who first thought of traveling beyond the Earth after watching episodes of *Star Trek* is nearly cliché. Yet the goal of good educators everywhere, at schools and museums, in talks, exercises, and exhibits is to open up minds.

Such experiences, behind closed eyelids, can be no less immersive and no less soul-shaking for coming to us in the unremarkable, quiet moments in our bedrooms or on the bus. They are insistent things—they compel us and carry us through years of study. Or they become companions to us, allowing us to explore the yawning recesses inside our own heads between the harsh realities of life.

As a planetary scientist working with spacecraft imagery and an astronomer specializing in public outreach, we happen to believe that the best kinds of experiences are those that combine both the physical world and the imagined one. Today, our robotic spacecraft are on the very edge of exploration. No human being has ever reclined on red sand and looked up to watch clouds scudding along in the Martian sky as the Curiosity rover has done. Nor have we parachuted beneath the Titanian clouds to land amid wet orange rocks, the ground giving off a puff of evaporating methane as we crunch through the crème-brûlée crust of the surface, as the Huygens probe has.

Instead, through the images they send back, in their sniffing of the air and tasting of the soil, our spacecraft tell us a small part of the stories of these places. It is then our imaginations that weave together these details with our own experience of the natural world here on Earth. In that sense, the other planets of our own solar system represent a perfect balance of the exotic and the familiar. Understandable on the human scale, yet undeniably unearthly.

Where spacecraft give us a window onto these places, it is our intention to pull you completely through that looking glass: to transport you and to encourage you to imagine yourself there.

Each chapter is divided into two parts. First, we will take you on a daydream to a place that we have explored in the past few decades through our robotic avatars. To help you imagine yourself in that place, we've been careful not to describe the protagonist of the daydreams except for their tools and actions. Next, we will share with you what we know in terms of the cutting-edge science of that place. It is that knowledge which informs

# DAYDREAMING IN THE SOLAR SYSTEM
## Book Map

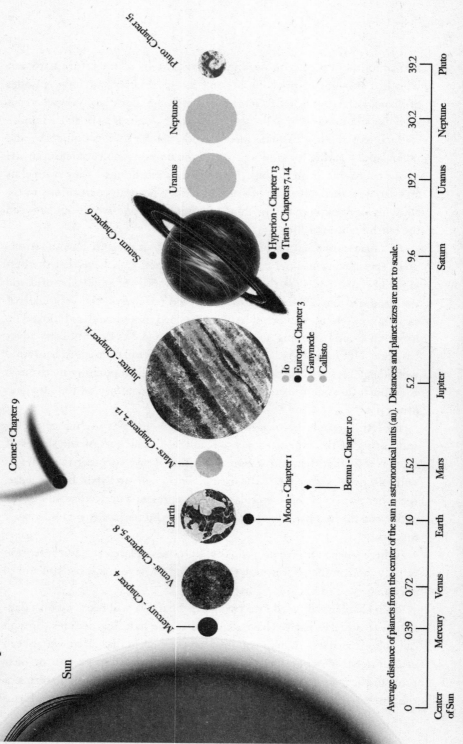

Sun

Mercury - Chapter 4

Venus - Chapters 5, 8

Earth

Moon - Chapter 1

Comet - Chapter 9

Mars - Chapters 2, 12

Bennu - Chapter 10

Jupiter - Chapter 11

- Io
- Europa - Chapter 3
- Ganymede
- Callisto

Saturn - Chapter 6

- Hyperion - Chapter 13
- Titan - Chapters 7, 14

Uranus

Neptune

Pluto - Chapter 15

Average distance of planets from the center of the sun in astronomical units (au). Distances and planet sizes are not to scale.

| Center of Sun | Mercury | Venus | Earth | Mars | Jupiter | Saturn | Uranus | Neptune | Pluto |
|---|---|---|---|---|---|---|---|---|---|
| 0 | 0.39 | 0.72 | 1.0 | 1.52 | 5.2 | 9.6 | 19.2 | 30.2 | 39.2 |

Figure 0.1
Welcome to our solar system! In this book you will explore environments on many of the planets along with asteroids, moons, comets, the rings of Saturn, and even a dwarf planet (Pluto). Use the map to the left as your guide while you read through the stories and the science that follows.

←————————————————————————————————

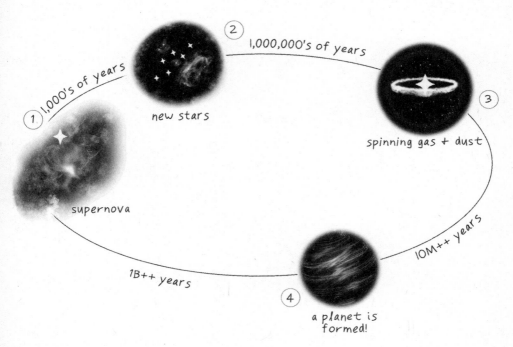

Figure 0.2
Solar systems like ours are formed when stars in or near nebulae explode at the end of their lives in an event called a supernova (1). The dust and gas traveling outward at high speed from the supernova compress the surrounding gas and dust, causing this material to mix and clump together. These clumps collapse under their own gravity to form new stars (2). Zooming in on those new stars, each is surrounded by a disk of gas and dust (3), which, over tens of millions (10M++) of years, becomes organized into new planets (4). Billions (1B++) of years later, the cycle begins again.

the daydream, allowing us to explore what it might be like to walk, play, sail, or fly there and to share some of the excitement and awe of that imagined experience with you.

If we've done our job right, by the end of this book you'll feel as if you have actually been to these places, even if only in your imagination.

Our first daydream keeps us close to home with a visit to our nearest celestial neighbor, the Moon . . .

GOLFING ON THE FAR SIDE OF THE MOON

The first thing that hits you in this place is the darkness. It's not absolute, of course. There are still stars up there, and a few planets are visible from where you are. You gaze upward and reflect while reclining on the fluffy, finely crushed rock material called regolith here on the far side of the Moon.

This place has the distinction of being the only part of a planet or moon where spacecraft eyes were needed to see what was out there. The Moon rotates at exactly the same rate as it orbits the Earth. As a result, the same side is always turned to face our home planet, a condition that scientists call "tidal locking." Therefore, no matter how powerful, a telescope located on Earth would never be able to show us the lunar far side.

Because of its unexplored nature, prior to spacecraft exploration, some called this place the "dark" side of the Moon. But even without a human to witness the passing of the days, they still pass as they have everywhere else on the Moon since the time of its formation. The only difference from a day on Earth is that those days last nearly a month here on the Moon.

Still, even if this side of the Moon is not bathed in perpetual darkness, the two-week nights are still different from even the darkest night on the Earth far from any city. Here, with no atmosphere, there is no sky glow from distant lights and not the faintest twinkle to any of the stars. It's much easier in this place to believe that those cold, crisp lights are each their own nuclear furnace, like our Sun, just much farther away.

Those stars provide no heat and barely any light. If you were to add them all up, they would total more than 100 million times less light than you would get from the Sun during the day here, or more than 1,700 times dimmer than the illumination of the full Moon as seen from Earth. Under such faint illumination you can barely make out your feet as you travel your nightly route! Frankly, humans were not built to walk over this ground unaided.

In Apollo images, the surface of the Moon appears blindingly bright. But that effect is just a contrast with the inky blackness of the space above the horizon. In reality, the lunar surface has the reflectivity of worn asphalt and its soft fluffiness from untold eons of micrometeorite impacts serves to soak up any angles that could help with orientation. You can prove this to yourself even on Earth by comparing the brightness of the full Moon just above the horizon to a brick or concrete structure just before darkness when both are illuminated by the same Sun at the same angle.

But what makes this a challenging environment for us makes it an excellent place for telescopes. The lack of atmosphere even means they can be used during the lunar day, if properly shielded from the Sun.

You swear you can see the astronomical instruments of the lunar far side observatory now as even darker silhouettes against the starscape above. It's as if they are dark constellations or nebulae, which hide from you the very universe they are meant to illuminate. Every so often, you'll see distorted mirror images of stars in their reflective surfaces, as if the sky itself has been moved and bent into new configurations.

Even without the dimness of the scene, the human eye has tremendous difficulty judging distance on this alien plain. Our stereo vision gives us helpful depth perception only out to a bit over five meters; beyond that, we typically rely on our experience of buildings and other objects to gauge scale. But here, a mound in the distance could be a molehill as easily as a mountain. Except, of course, the Moon has no moles. At larger distances, the eye is tricked by the closeness of the horizon. Sitting atop a hill on the Moon of the same height as one on Earth, the objects on the lunar horizon are only half as far away as they would be at home. But your brain doesn't know that, at least not on a visceral level.

Clearly, this is not a place that is kind to human bodies or to human senses. You find it vaguely amusing that this reality exists even though the Moon is our nearest neighbor. Indeed, the very rocks under your feet were literally once a part of the Earth itself. They only became part of an unusually large and close-orbiting satellite in the fiery aftermath of a titanic celestial traffic accident during the early history of the solar system. At that time, not long after the Earth had formed, a Mars-sized planet collided with our home world, sending enough material skyward to create the Moon. That truth is written in the very chemical makeup of these soils—one of the great and surprising discoveries of the Apollo missions.

Those Apollo explorers from long ago did not stay on the Moon overnight, and for good reason. During the day, temperatures can reach over

100°C, but at least you can find shade or bring your own. At night, without the protection of an atmosphere, it gets exceptionally cold: 130°C below zero is common. In some places that are protected from the heat of the day, temperatures can get to 200°C below zero or even colder.

The coldest places, located near the poles, are called "permanently shadowed regions." Though there is no dark side of the Moon, there are dark patches! These places only ever see the light of the stars. Here, it is cold enough that any water or volatile gases that waft to the surface will freeze and stick around nearly forever.

Paired with those permanently shadowed regions are ridges of permanent sunlight. These are fantastic places for observing the Sun. Your first job here on the Moon was at one of those observatories. You remember well hiking down into the permanent shadows to harvest the ancient water to support the lab. Down in that strange bowl of darkness, with a blinding light bouncing off the high cliff faces, the illumination was surreal.

On certain days, you could even make out the full disk of the Earth, looming on the horizon. At more than 13 times larger and three times as reflective as the Moon, a "full Earth" provides nearly 40 times as much light as the full Moon, and the colors and features are immediately recognizable.

It truly is a sight to behold.

But beyond the math, you were rendered speechless by the sight of humanity's home, just like the astronauts of Apollo 8 were in the 1960s: a fragile bauble suspended over the horizon where every single human being ever in existence, indeed all the life that was known in the universe, resides. You realize that it wouldn't take much on a cosmic scale to be looking back at merely a larger version of the barren, pockmarked wasteland that is the Moon rather than our oceanic garden planet.

At that solar research station, the return from the shadowed deeps was a hike from perpetual darkness into perpetual light. The experience was transcendent but felt somehow artificial. Eventually, you instead decided that you preferred the far-side observatory to the endless day of those ridges. When an opening came up there, you jumped at the chance.

You've found camaraderie here, of a sort you haven't experienced since you worked at McMurdo Station in Antarctica, many years ago. There's something about the isolation and the extreme conditions that bond you and your fellow technicians.

That and the fact that you and your friends on the staff can take full advantage of the Moon's low gravity for some truly unique invented activities. When the engineers built this place, several extra empty telescope enclosures were included to allow later expansion. For now, let's just say it's amazing what new sports you can come up with in a cavernous space where a typical jump launches you three meters off the ground and hangs you in the air for four seconds! To say nothing of what you can do with an improvised trampoline or a ball. Truly, it is only here on the Moon where the body can fully inhabit our heritage from creatures who were at home swinging in a forest canopy.

You and your colleagues almost never play the same game twice. Of course, you still must be careful—with the same mass and inertia, it can be painful to run into one another or to fall from a height.

But sometimes, all you want is a simpler pleasure. A pleasure that even the Apollo astronauts knew. The telescopes are laid out in a configuration that looks like a large letter Y on the landscape. This allows the telescopes to perform a trick to get especially clear images that the astronomers call interferometry. So, you head over to a slightly raised spot you frequent just beyond the bottom of the Y.

As you look out at the surface of the Moon beyond the array, you pick up a golf club that you have left here. The club is unchanged since you dropped it, and it will look the same for millennia to come—dust deposition is very slow on the Moon. You know most dust comes from a process called impact gardening, in which small micro-meteorites smash into the surface, sending particles on parabolic trajectories. The surface of the Moon is so broken up that this process mostly just moves material around, creating soft edges and hills over time.

But one of the fascinating consequences is that there is always some dust in transit above the lunar surface. Glancing off to one side, you know that sunrise is near because you can see just a faint trace of "lunar horizon glow," the light reflected by all those bits of dust while they remain "air"-borne. The light looks so stable, you'd never imagine that it was the product of such a random process.

From a pack slung across your back, you withdraw some special golf balls. These balls are specially made with a light inside, so you can see your shot travel in the darkness.

You turn on a ball and set it on your tee. As you prepare your swing, you are thankful for the modern technology that gives you a much more flexible spacesuit than previous explorers enjoyed. Alan Sheppard would be jealous. The Moon's low gravity means that you once managed to send a ball up to 2,000 yards. But tonight's not about breaking records.

You swing back, and as you let go, you feel the club strike the ball through the vibrations that pass up the club and into your hands. It's a satisfying feeling that tells you that you connected well even though there is no atmosphere to transmit that ping of contact.

As you watch the ball recede, you think to yourself, Could there be a purer demonstration of the dance between motion and gravity? The ball flies so high and hangs so long that you lose track of it. Eventually, for all the world, it looks just like any other star. Not to worry, you'll collect the balls you drive during the lunar day—it's easy to make out such brightly colored objects against the gray background here.

You strike a few more before heading in, your rounds completed for the night.

As you pack up and prepare to return to the common building, you reflect on the strangeness of this place. Here, you come face to face with the unimaginable magnitude of cosmic scales and processes. Though the human mind can't hope to truly take it all in, you find you don't need to look away to appreciate the grandeur of this "magnificent desolation."

Instead, you spare one more gaze out at the landscape in wonder before setting your mind and body, once more, to the task at hand.

★★

The Moon is the most natural place to start our daydreaming in the solar system, because it's the only place humans have experienced firsthand. In the 1960s and '70s, NASA astronauts walked, ran, jumped, fell, and even golfed on the Moon. Having watched those people experience such an alien place, we hope this made it easier to metaphorically step into those space boots.

Though we call it the Moon, officially, the Moon is a *natural satellite* of Earth.[1] Of course, in conversation, we often use the word "moon" to refer to natural satellites. For example, Phobos is a moon of Mars.

## WHAT'S THE DEAL WITH SURFACE GRAVITY?

The sensation of gravity we experience daily here on Earth is built so deeply into the human experience that we can forget it's even there. Yet, in everything we do, every action we take, we compensate for gravity. This sensation of gravity, and its strength, is determined by our specific scenario here on Earth. We'll come back to that in a minute.

Gravity is a fundamental force of the universe, and like any force, it will act to accelerate anything with mass.[2] When things fall on Earth, they don't fall at a constant speed; they continually accelerate toward the ground.[3] One way of conceptualizing the strength of a planet's gravitational field is to measure how much something accelerates when free falling within it. On Earth, that acceleration is about 10 m/s². This means that as an object is freely falling toward the Earth, it will increase its speed by 10 m/s every second until some other force acts on it (such as wind resistance or the ground).[4] The force created by the Earth's gravity is defined as 1 g (pronounced *one gee*).

So why does Earth's gravity accelerate a falling object at exactly that number? What determines the magnitude of that acceleration? We can hear you screaming "the mass of the Earth of course!" and you'd be right, but it's not just the mass of the Earth; it's also the *size* (i.e., radius) of the Earth that matters. How does this work?

Acceleration due to gravity of something falling toward any massive object is calculated using the following equation:

$$g = \frac{GM}{r^2},$$

where $M$ is the mass of the object and $r$ is the radius of the object. The value $G$ is the universal gravitational constant. Earth's mass is roughly $6 \times 10^{24}$ kg. That's a 6 with 24 zeros after it, or 6,000,000,000,000,000,000,000,000 kg. Earth's radius is roughly 6,370 km.[5] If you study the equation for a minute, you will see that the gravity we feel, $g$, depends on two factors. The first factor is mass; clearly, the more massive a planet, the greater the gravitational pull. The second factor is that planet's radius. If the Earth had the same mass but was physically smaller in size, then the value of $g$ would increase. This makes sense because you are physically closer to all the mass that is pulling on you. Alternatively, if Earth had the same mass but was physically larger in size, the value of $g$ would decrease.

So, when calculating the strength of surface gravity of some asteroid, moon, comet, or planet, both the mass and the size of the object matter greatly. We'll see this come up again and again throughout our exploration of the solar system.

Let's turn our attention to the Moon. It has a mass of $7 \times 10^{22}$ kg and a radius of 1,737 km.[6] If you calculate the value of $g$ on the Moon using the above equation, you get an acceleration of 1.6 m/s$^2$. That's about six times less acceleration than the amount created by Earth's gravity; we could say that the Moon's surface gravity is 0.16 $g$.

In a gravitational field that accelerates you six times less than on Earth, you could jump much higher using the same jump force, and you would take longer to fall back to the ground.

But there's an important factor to consider, as noted by our daydreamer walking around on the lunar far side. That is, while the strength of gravity is less, meaning you accelerate more slowly toward the ground, you still have the same mass. Since you have the same mass, the amount of inertia you generate when you're walking, running, or hopping along the surface is the same as it would be on Earth. Another way of saying this would be that if you were running along the surface of the Moon and crashed into a rock, you would feel the same pain as if you were running along the surface of the Earth at the same speed and crashed into a rock.

## OBSERVING THE SKY FROM UNDER A TURBULENT FLUID

Astronomers have a strong dislike for both clouds and city lights. Clouds can roll in and out with very little notice, making it difficult or impossible to collect light from stars, galaxies, or anything else interesting out there in space. However, there are some astronomers who don't care about clouds, for example, radio astronomers. A radio astronomer is a scientist who specializes in collecting radio light coming from the universe. Stars, galaxies, neutron stars, and many other astronomical objects shine in radio light. This light can't be sensed by the human eye, but we can observe it with special instruments. Radio light goes straight through clouds, thankfully, considering we use radio light for communications. You wouldn't want a call to drop or a message not to send because a cloud got in the way. But if you are trying to observe the sky in optical light, then clouds are clearly going to give you trouble.

But clouds aren't the only barrier our atmosphere creates for earthly observers. Just the very fact that we have an atmosphere in the first place can make observing difficult. This is because there are many things suspended in it, like small dust or particles. This becomes very apparent near or in a large city. Light from the streetlights, buildings, houses, cars, and other sources shines in all directions, including up. And while most of that wasted light just happily goes through the atmosphere, some of it bounces off those tiny little dust particles floating around. As it bounces around the dusty atmosphere, it creates a glow that we call light pollution.[7] This is why astronomers try to put their telescopes in the most remote places possible, like deserts, the tops of mountains, or other places far from human habitation.

But even if you were able to get to the most remote place on Earth with no clouds, or even dust, the atmosphere would still have an impact on your observing. This is because Earth's atmosphere is essentially a big layer of fluid in between us and space, and that fluid has turbulence. Think about it sort of like swimming to the bottom of a pool and looking up toward the surface. From your perspective below that layer of water, anything above the surface is distorted. The images are often wobbly and shaky due to the bulk motion of the water in the swimming pool, and the effect gets even worse if someone ripples or splashes the water.

Earth's atmosphere behaves in a similar way: optical light from the universe travels through our turbulent atmosphere, which can ever so slightly bend the incoming light. We see the result of this effect in the apparent twinkling of stars.[8] This is an effect we astronomers call *scintillation*. It certainly makes the night sky look pretty, but to an astronomer, the scintillation reduces the precision of our measurements. So, what's the solution? The simplest way to fix the issue: get above the atmosphere! Just like when you swim back to the surface from the bottom of the pool, and you can see everything clearly above the water, if we put our telescopes above the atmosphere, then we don't have to worry about its muddying effects. This is the logic behind putting the Hubble Space Telescope and the James Webb Space Telescope in space.

## TIDAL LOCKING

The concept of tidal locking can be confounding, yet we've been experiencing it our entire lives. Humanity has been staring up at the Moon since time immemorial, and that entire time we have always been looking at the same side of the Moon. We all know it's a sphere, and therefore has a full 360 degrees of surface area, yet we only will ever see the one side, regardless of where you are on Earth or at what time you look. We've become so familiar with the side of the Moon that faces us that we have created numerous images we often point out to each other, like "the man in the Moon" or "the Moon rabbit."

So why can we see only one side of the spherical Moon from Earth? You might find yourself wanting to say it's because the Moon doesn't spin, but this would be wrong (and a common misconception). The Moon is very much spinning on its north-south axis, just like the Earth spins on its north-south axis. So, if the Moon is spinning, then why do we see only one side? The reason is because the time it takes for the Moon to complete one full rotation on its north-south axis is exactly equal to the time it takes for the Moon to complete one orbit around the Earth, which is about 27.5 Earth days. This is called a synchronous orbit and is illustrated in figure 1.1.

In our daydream, you learned some of the benefits of placing observatories on the far side of the Moon. In figure 1.1, an observatory is constructed in just such a location and is shown to always be facing away from Earth. As the Moon orbits, the observatory periodically experiences sunrise, sunset, day, and night, but never comes into view of the inhabitants of Earth.

The process by which the Moon's orbit became synchronized is called tidal locking. When the Moon formed from the aftermath of a catastrophic planetary collision, it was originally spinning much faster, such that the Earth would have been able to see all sides of the Moon (though there was no one around to observe it!). Slowly, through time, tidal forces acted to spin down the Moon so that its spin around its axis matched its orbit around the Earth.

You may recognize the root of the word "tidal" from the tides of the sea, the phenomenon of ocean waters gradually receding and returning once every 12 hours. These periodic tides are caused by the Moon's gravity pulling on the Earth and creating bulges in our oceans, which are freer to respond to the gravity than the solid rock of Earth. The Earth, however, is continually spinning despite these bulges in the ocean, which creates friction. The result: the Moon's tidal forces are acting to slow down the Earth's

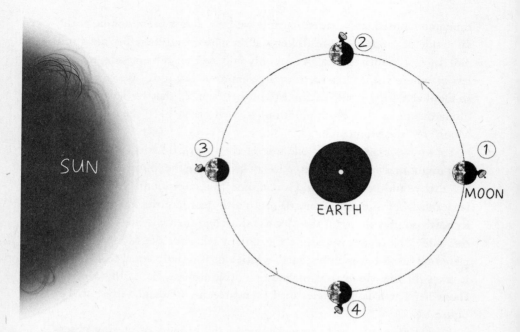

Figure 1.1
A view looking down on the north poles of the Moon and the Earth. As the Moon orbits the Earth, it keeps the same hemisphere facing toward us. When the Moon is on the opposite side of the Earth from the Sun (1), the side facing away from the Earth (the far side) is also the dark side of the Moon. But as the Moon travels along its orbit, the Sun will rise on our observatory (2) and eventually the entire far side will be in sunlight (3) before sunset at the observatory (4) begins the cycle again.

spin, and thereby elongating our day. This is called "spinning down." But don't worry; while our day is getting longer, it is only fractions of a second every century.

The Moon isn't the only celestial object with tidal forces; the Earth also acts on the Moon in the exact same way, but the Earth is much more massive, with much more effective tidal forces. The Earth's tidal forces have gradually slowed down the Moon until its spin is exactly synchronous with its orbit. Surprisingly, this might have taken as little as a few thousand years after the Moon formed.

Since the Moon is tidally locked to us, that means the only way we can ever see the far side of the Moon is if we fly a spacecraft there. Indeed, the first time humanity saw the far side was in 1959 when the Soviet Union flew Luna 3 around the Moon and took pictures for us all to enjoy.

Eventually, many billions of years from now, both the Earth and the Moon will be tidally locked to each other, showing each other the same face for eternity.

## MEASURING DISTANCES IN THE SOLAR SYSTEM

When we talk about the Moon, we're close enough to Earth that measuring distance in kilometers still seems appropriate. The Moon is, on average, 384,400 km away. That is still very far for humans, though; traveling at highway speeds of 100 km/h, it would take 150 days to traverse that distance. Luckily, our rockets can make the trip in just a few days.

We say the moon is "on average" 384,400 km away, because it does not maintain the same distance away from us. Like all orbits in the solar system, the Moon's orbit is an ellipse, not a perfect circle. At its closest to us, the Moon is just 363,300 km away (a.k.a. perigee); at its farthest it's 405,500 km (a.k.a. apogee).[9] Each time the Moon orbits us, it goes from its closest to farthest and back again.

But once we leave the relative proximity of the Moon, kilometers as units of distance become cumbersome. Jupiter is 778,479,000 km from the Sun. Saturn is 1,432,041,000 km from the Sun. Pluto is 5,869,656,000 km from the Sun. You see? Just like you wouldn't want to use meters or centimeters to measure the distance traveled on a road trip, we don't want to use kilometers to measure distances in the solar system.

To address this issue, space scientists developed the *astronomical unit* (au), a unit of distance that helps us better manage and picture the vastness of our solar system. The definition of one astronomical unit is the approximate average distance from the Earth to the Sun. Earth, on average, is about 149,597,870.7 km from the Sun, which we can now call 1 astronomical unit.[10]

The power of the astronomical unit is twofold: not only is the number 1 so much easier to say than 149,597,870.7, but with this unit, for every distance we measure, we are comparing with the Earth-Sun distance. For example, the destination of our next daydream is Mars, which orbits the Sun at an average distance of 227,956,000 km. We can convert this to astronomical units using the definition above:

$$\frac{227,965,000 \text{ km}}{149,597,870.7 \dfrac{\text{km}}{\text{au}}} = 1.52 \text{ au}$$

Mars is about 1.5 au from the Sun. That's easy to say, and we now know that Mars is about 1.5 times farther from the Sun than we are. A fun challenge would be to figure out what the Earth-Moon distance is when expressed in astronomical units instead of kilometers.

# RAPPELLING INTO VALLES MARINERIS ON MARS

You left your vehicle behind at midday so that you could hike far enough in daylight that it would be over the horizon from your perspective by the time you made camp. Why make the coming descent harder than it needs to be? Quite simply, you wanted this moment to belong just to yourself and the landscape: sunset on the rim of Valles Marineris, the largest canyon in the solar system.

And what a moment it is.

You dangle your feet over the edge of this spur into the Coprates section of Marineris and drink in the view as the Sun descends into the canyon on this balmy austral summer's day. From your perspective, the floor of the canyon is almost 10 km beneath you. Across the way, from your perspective and nearly 100 km distant from you, the wall of the other side of the canyon rises up again to your level, smoky through the suspended dust.

That far wall isn't even the other side of this massive gorge; instead, it's just a finger of land called the Coprates mountains protruding into the center of the great gap. There's a whole other tributary of Marineris on the other side, hidden from view around the curve of the planet. At its widest, Marineris is nearly 600 km across.

As a result, not only is the scale of Marineris difficult to hold in your mind, but you also literally cannot appreciate its entire extent from any point on the surface. Generously, Mars holds back and presents only what you can comprehend. Each view is a frame in a movie that evolves as you travel the nearly 4,000-km length of this feature from Noctis Labyrinthus in the west to the Aurorae Chaos in the east. You take a moment to consider that this distance is farther than the distance between the surface and the center of the planet.

By contrast, the Earth's Grand Canyon is a mere 29 km wide and only about 2 km deep, extending for less than 500 km. A sight to make a human feel small, certainly, but a feature that would be lost in insignificance on Mars. Marineris alone could enclose hundreds of times the volume.

The sky darkens and shifts to cold hues in the long light. Though it looks pure white in photographs because of the exposure time, to human eyes the solar disk itself turns a blue-white —you are looking through so much of the atmosphere at this angle that it is as if the Sun has been exsanguinated. On Earth, volcanic ash turns sunsets to a deep red color, scattering away the blue light. But Mars's dust is different, and the particles here combine to scatter away the sun's red light into the rest of the sky, bleeding the Sun dry just before the day ends. On especially dusty days, the Sun is lost entirely even before it reaches the horizon, dissolving slowly into a thick haze.

Slowly, the stars come out and the vault of the heavens is revealed. But on Mars there is an interloper. A bright evening double star appears in the west, following the Sun. It's the Earth and the Moon. Before coming to Mars, you wondered whether you would be able to see both together or whether they would be blended into a single bright point. Now you realize it's easy to pick them apart. When Mars is closest to the Earth, they can be separated in the sky by more than a quarter of a degree, or about half the width of the full Moon as seen from the Earth.

Of course, Earth gets the better of this arrangement. Someone back at home would observe spectacular views of the fully lit hemisphere of Mars during these oppositions. But from Mars, both the Earth and Moon would appear as thin crescents quite near to the Sun in the sky as they followed that copper coin through the daytime. That geometry makes it much harder for a casual Martian observer to get a good view unless they have some effective sunglasses or can time their observing right near either sunrise or sunset, when the bulk of the planet blots out the Sun and if the atmosphere is relatively clear of dust.

Turning your eyes from the sky, you start going through your gear in anticipation of the next day. You lay out everything and test it. By and large, it is a typical climber's kit—something recognizable to even ancient mountaineers on Earth. You have ropes and mechanical ascenders, pitons, and other paraphernalia, even an ice axe and ice screws in case the materials of the cliff face change on your way down.

But there are a few tweaks to make these tools more appropriate to Mars. For one, all the equipment seems surprisingly slender and light—the advantages of working in the one third gee on this planet. Additionally, your rope is equipped with an electrostatic clasp. In this way, you can recover your own rope after you have rappelled down a course. That saves you from having to bring too much weight along with you. Indeed, the whole kit fits neatly in a large climbing pack.

As you lay down for the night, you go over the route in your mind. The first course is steep until you get to a ridgeline. You can then follow the ridgeline to two false peaks before the terrain falls away steeply near the bottom. Right where the slope ends, there is a small unnamed crater that you can use to guide your pathway. All told, it's just over 9 km down in a little over 20 km as the crow flies. With an average slope of just a bit over 20 degrees, there will be a combination of nearly walkable slopes and sheer cliffs.

This part of Coprates is as steep as the cliffs that encircle Olympus Mons, but considerably higher. When you throw in the weaker gravity with such colossal ascents, Mars is clearly a climber's paradise. Plus, going down is much faster than going up. Where it might take up to a couple of months to climb this wall of rock, you anticipate getting down in no more than three days. It's for this reason that you have foregone a pressure tent. That will help you to move quickly, light upon the land. But you anticipate that you will look forward to a shower and a stretch with increasing anticipation as you descend.

You smile as you close your eyes. Your last thought before drifting off is that it must look strange to see a person lying on the ground on their back in the middle of the night. Anyone who came upon the scene might be tempted to ask what misfortune you encountered. But, in a spacesuit, you're effectively wearing your own tent—so why not lie down on the regolith and sleep under the stars?

The next morning, your suit wakes you before sunrise. Gathering your kit and walking up to the edge, you are greeted by a strange sight. As the sky begins to take on color, you can see that the canyon system is filled by a sea of fog. The earliest robotic explorers reported seeing the same phenomenon. But when later spacecraft came back to map Mars from orbit, they chose to observe the planet in the early afternoon when the surface was least likely to be obscured by clouds. For a time, the morning fogs were forgotten.

Examining the fog more closely, it seems to rise to the same level on the opposite side, as if there was a fluffy sea filling the canyons. The upper surface even has wispy waves. You know that this fog will dissipate later in the morning and that there is not enough moisture to make the slope slick and cause a hazard. But you realize that you may well start the next two days in a cloud.

However, at the top of the canyon, conditions are clear, so you are ready to set out. Your first pitch will take you from just below the 4,400-meter level above the zero elevation line, called the datum, all the way down to just below the 1,800-meter level at a false summit protruding out into the canyon. As the crow flies, this will be 7,760 meters from the edge for an average slope of just 18 degrees.

The very start is one of the places where rappelling is required. Here, the edge of the canyon has eroded in a wide, semicircular bowl shape. Initially, the slope is very steep but there are plenty of cracks and footholds and you find the going easy. However, the angle eventually eases until it becomes less than the angle of repose of granular materials, about 30 degrees on Mars. Here you are in a sea of broken rocks and soil known as talus. Unfortunately, this means that there is little stable material. This slows your movement tremendously as you scrabble along to find outcroppings and boulders that you judge to be large enough and stable enough to serve as places to fix your lines amid all the loose material.

By the time you reach the 3,300-meter level, the ridgeline starts to protrude assertively from the debris. You find that in some places you can walk directly along the ridge and can make excellent time. However, in other places, the ridge breaks up into a blocky, stepped structure that requires checking your harness and rigging lines even for fairly small drops. But the work is pleasant and the problem-solving nature of picking a course down the slope fills your day with a productive focus.

Many descents into canyons are like this. Just like in a terrestrial canyon, the wall here is made up of layers of rock that were deposited eons ago into horizontal layers and have since eroded into the forms you see. It doesn't matter whether those layers are volcanic rocks or sedimentary rocks; having formed at different times and in different ways, they will all have different strengths. That means that some will crumble away easily, leaving gentle slopes. Other, stronger layers will stay intact. This leads to a sheer drop until the erosion of an underlying layer completely undermines the layer. At that point, a block falls away to the floor below and the slope continues its inexorable retreat from the void.

On your breaks, you take the time to gaze out at the canyon and to appreciate the way that the light interacts with the landscape. The canyon runs roughly along a line of latitude, so in the morning, with light coming from the east, the Sun nearly rises out of the canyon itself in a

Stonehenge-like alignment. This casts long shadows from every crack and crevice in the canyon walls. Gradually, the Sun climbs into the sky and the far wall descends into darkness. With the flat illumination of scattered light, you can see contrasts in color and texture clearly. As you glance across at the opposing slope of the Coprates mountains, you can see layers. There are dark marker beds that you can follow for miles.

Unlike at Olympus, here you can see the entire height of the matching wall. Marking this terrain as incredibly ancient are the impact craters that mar the surface. Your wall must have similar features, but they are hidden by the terrain right in front of you and underneath. Whether you are climbing or rappelling, it often feels as if you are an ant traveling along a tilted world. It's a perspective that gets you up close with your surroundings, with every nook and cranny. But it also hides the bigger picture.

By the end of the day, you arrive at your second camp dusty, but having made remarkable progress. Quite frankly, if this were Earth, you could not have succeeded in traveling this far or descending this much distance so quickly. But on Mars the rules are different. As you eat a supper of relatively tasteless but nourishing rations through a straw in your suit, you think a little bit about the different elevation levels and how these are defined. As a result of where you are climbing, you are starting well above the datum, and will be descending almost an identical distance below this somewhat arbitrary marking line.

The last thing you do before you end your first day of climbing is to clip your harness firmly into a rocky outcrop nearby. It wouldn't do to go sleepwalking over the side in the night!

The next day dawns in a fog, as you expected, and so you use the time until it burns off to plot out your second day's route. Where your first day was oriented south by southwest, today you are moving with the topography and following another ridgeline that departs from this point to the left along a due south heading. Again, today you will cover a similar height drop and horizontal distance, but you anticipate the route to be easier since it will be all along the ridgeline. The goal here is to set up a spectacular day three when you'll drop nearly four kilometers to the floor on some of the steepest terrain Mars has to offer.

Once the fog dissipates, you take a minute to look across the gap. Like the scarps surrounding Olympus Mons, Marineris owes its existence to tectonic forces. In the case of Olympus, the incredible bulk of the mountain pushed down on its center while raising its edges. Eventually, the entire shield cracked off at nearly the same distance from the lava source, creating the circular scarp we see today.

The formation of Marineris also released stresses, but these built up over the whole of the Tharsis volcanic province. Here, this enormous stack of lavas, thousands of kilometers across, slowly relaxed into the planet, pushing aside mantle material with their bulk. Just like a cheesecake that has been cooked too long in the oven and whose surface is no longer pliable, this relaxation caused radial cracks to appear as the layer fell. However, instead of many cracks, Tharsis produced just one massive "graben": Valles Marineris.

Not all canyons have this history. The Earth's Grand Canyon, by contrast, was formed by a river, eroding rocks over time. Yet, some do form in the Martian way: the canyonlands to the north of the Grand Canyon are a great example. Here, rock trying to stretch was broken into parallel channels where the rock was fractured and fell to produce a flat valley floor and steep walls.

But these days, the titanic forces that built this landscape are quiet and so you can continue your descent into Mars without too much fear—that is, aside from the occasional rockfall. As a boulder whizzes by off to your left you are reminded that an intensity of erosion that is insignificant to a huge feature like this canyon can still be lethal to a single human climber. Suitably chastened by nature's roll of the dice, you continue toward camp three.

Another spectacular sunset and night. At this point you are below the datum, having descended more than five kilometers. The wall opposite you now visibly towers above your level. Outside, the atmospheric pressure is more than double what you experienced on the rim. As you recline on the surface, you wonder if that faint whistling you hear is the sound of wind passing by your suit, or just the air exchanger?

The last day dawns especially foggy, yet you have no choice but to proceed. This will be your longest pitch over the steepest terrain. Indeed, today you anticipate that you won't finish until well after sunset. It is here, just beyond camp three, that the ridgeline dissolves into sheer drops, eventually running out into gullies.

You attach your safety line and begin.

It isn't long before you reach the first drop of the day. You attach your line and proceed over the edge. As you descend into the void, you realize that this is an overhang with the wall arching back away from you. In the fog, that wall becomes fuzzier and fuzzier until it disappears from view entirely. You find yourself in the strange situation of being suspended in a cloud by a thin rope. In the milky light, you consider that you don't know how far this overhanging layer extends. If this layer is too thick, you'll run out of rope and will need to climb back up and find another pitch. But for now there is nothing for it but to keep going.

As you descend, suddenly you are out of the fog and into clear, thin air. Looking upward, now it seems as though you are suspended from a cloud by your rope. The visual gag makes you chuckle. Looking horizontally, you get your bearings and see the floor of the canyon stretching out under the flat lighting below the fog. It's as though you have discovered a vast interior world, bounded by the cliffs on all sides between the canyon floor and the clouds. Looking down, you are relieved to discover a perfectly flat bench of rock that you will reach before you run out of line. Looking back at the cavity under the overhang you see ice, apparently frozen in the act of seeping out of the rock. Perhaps this is what has caused the increased erosion in this place?

You take a close look at the ice, faintly blue in color, translucent, and hard as concrete. This water has likely been sitting here, kilometers below the surface, since the end of the Noachian period, billions of years ago. It's likely no one knows this outcropping—radar would not have been able to detect such a thin layer so far below the rim's surface. You'll have to note the location for some scientist friends of yours when you get back. Skillfully, you chip off a few pieces and tuck them into the water pack of your suit. Soon enough you'll be drinking ancient Martian water, and a few of those molecules will become a part of you permanently.

Once you retrieve your rope, you sit with your feet over the edge, awestruck as you have been so many times already on this journey. As you drink the ancient water, you drink in the timeless scene out over the ledge. But you can't stay here in this ethereal world; you have places to be and "miles to go before you sleep." So, you take on the next pitch and descend again.

The rest of the slope is mostly repetition, but eventually you reach the foothills below in a tumble of building-sized boulders. In some cases, you pick your way over top; in others, you travel in between, down through uncountable slots and crevasses. At first you are traveling through the deep shadows of the late afternoon, but eventually night falls and the gaps between the boulders frame small snatches of sky. Here a selection of stars, there a galaxy or nebula, or a constellation—they are the same on Mars as those you remember from your youth on Earth.

Then, suddenly, you are out at the bottom, on the canyon floor. After so much verticality in the last three days, the flat horizontality of the surface seems wrong. In the distance, you are relieved to see your rover—it took the long road to meet you here, traveling thousands of kilometers autonomously to this rendezvous. You will be happy to get out of this suit.

Exhausted, you fall to your knees. Around you, the ground is covered with sand, and you wonder whether perhaps Marineris did once hold water inside its massive channel. Perhaps you arrived at some forgotten beach?

In the cold air above you, Mars's daily cycle grinds ever onward. You direct your flashlight upward and are rewarded by a million glints in the beam cutting through what looks otherwise like clear air. Diamond dust, the atmospheric scientists call it. Down here where the atmosphere is thick and can hold more water, the particles form faster and larger. This allows them to become individually visible, though still only tens of microns large, eventually tumbling to the ground like snow.

Switching off your suit's lights, you walk to your rover through a swarm of sparkles as the diamond dust catches shards of starlight, the double star of the Earth and Moon overhead.

★★

It's fitting to make our first step out beyond the Moon to the infamous red planet, even if it is not the closest planet to us. We chose Mars because, as we leave the relative familiarity and safety of our Earth-Moon harbor and head for open space, Mars is the next most familiar place. While Venus is closer to us (and the first planet to which we ever sent a robotic spacecraft), Mars has long captured the imagination of our sciences, engineering, and society. At 1.5 au, Mars is roughly 1.5 times farther from the Sun than we are. Though, just to be clear, that doesn't mean that Earth is 0.5 au from Mars. Both Earth and Mars are on orbits around the Sun. Sure, at our closest we're about 0.5 au apart, but if Mars was completely opposite the Sun from us, on the other side of its orbit from our point of view, there would be a full 2.5 au between us on a straight line.[1]

## EVERYTHING IS BIGGER ON MARS

Mars has the solar system's tallest volcano, Olympus Mons, with a height of about 22 km. However, where is that being measured from? Olympus Mons is standing 22 km above . . . what? On Earth, the level of the ocean is relatively flat across the planet and makes an excellent "zero"—or datum—elevation. Earth's tallest peak, Mount Everest, is almost 9 km above sea level. But on Mars, there is no ocean. When confronted by this problem, early spacecraft scientists considered different options. Perhaps the lowest point, for instance, should be zero? But instead, it was decided to establish a datum level around atmospheric pressure, specifically, the triple-point pressure of water. Like any atmosphere, the closer you are to the ground the higher the pressure, and the higher your altitude the lower the pressure. The datum of Mars was chosen to be the altitude at which the atmosphere pressure equals 6.1 pascals (Pa), or about 0.6 percent of the surface pressure on the Earth. This corresponds to the triple-point pressure of water, where $H_2O$ can exist as either a solid, liquid, or gas.[2] Mars also has the longest and deepest canyon system, Valles Marineris. Sometimes it just seems as though everything is bigger on Mars! But why is that?

Regardless of your location in the solar system, the ability to build a bigger mountain or a deeper valley depends on the strength of the materials (i.e., rocks, ice) you use and the ability of gravity to even out any differences over time. These two forces are always in tension on planets. For instance, planets, dwarf planets, and some moons are round because they are big enough that gravity wins against the strength of rock, crushing and breaking the rocks and drawing the planets into spheres. In contrast, asteroids and comets are both smaller and less massive, allowing the strength of the rock and/or ice to win out over gravity, giving them strange, irregular shapes.

The continuous tension between the strength of gravity and the strength of the building materials affects not just the overall shape of a planet, but also how large a geological feature can get. Beginning at the surface, as you descend into a planet, the pressure from all the rocks overhead pushes down on the rocks in the layers below. The farther down you go, the more rock is overhead and the more the pressure builds. Finally, starting from a few kilometers to a few tens of kilometers beneath the surface, depending on the planet, the pressure generated by all that rock being piled up exceeds the ability of the rock to support the weight. For most rocks, this happens at about 100 megapascals (MPa), or the level at which there are 10,000 tons per square meter overhead in a 10 m/s$^2$ gravitational field.[3] Increase the pressure beyond this point and the rock will crumble or move rather than support that weight. And if the rock is hot enough, it can even flow in the way taffy can when you pull on it.

When a planet is building a mountain, this sets a limit to height. If you build a mountain so high that the rock beneath is crushed, your mountain will slowly sink into the planet. Similarly, if you dig a very deep and very wide hole, like a canyon, the pressure from the canyon walls will cause the canyon floor to rise.

Planetary scientists call this process *relaxation*, and it doesn't happen quickly; it takes time for the rock to relax. For example, between 100,000 and 10,000 years ago, large portions of Canada (and other parts of the world) were covered in sheets of ice many kilometers thick; this is known as the Last Glacial Period. During this ice age, the continents were pushed down into the Earth slowly under the weight of the accumulating ice, and it took thousands of years for the land to rebound.[4] In the case of volcanoes, it can mean many millions of years or longer for a built-up volcanic mountain to come to its final compensated height, like an iceberg floating in the crust.

The time it takes for relaxation to take place is affected primarily by two variables: the type of material and the strength of the gravity field. If you play with these two variables, you can get some interesting results. For instance, in the outer solar system, there are many moons made mostly of ice (we'll visit a few!). Ice cannot bear as much pressure as rock does, crushing more quickly under a smaller mass above. Therefore, ice flows and relaxes faster. As a result, the surfaces of the moons of Jupiter are littered with completely relaxed craters that are nearly flat. If you lower the gravity, say, by going to Mars, the process happens more slowly. Mars is almost 10 times less massive than Earth, and about half the size (don't forget how we calculated $g$!). This makes Mars's surface gravity about 3.6 m/s$^2$, or almost three times less than Earth's.[5] Since there is less gravity pushing down on the rocks than on Earth, you can stack more material up to larger heights without relaxation being forced to compensate.

It turns out that of all the rocky worlds in our solar system that are, or have been, active enough to build mountains, Mars has the lowest gravity. That means it's going to have the tallest peaks.

## THARSIS: A VOLCANIC PLATEAU

Not only is Mars home to the tallest mountains and the deepest canyon, but it is also home to the largest volcanic province in the solar system: Tharsis.[6]

Tharsis, sometimes called the Tharsis Bulge or the Tharsis Rise, is an elevated plateau region that is more than 5,000 km across and, on average, about 7 km above the datum. This is a geological formation similar in size to a continent on Earth; in fact, the farthest distance from east to west in Canada is about the same size as the Tharsis plateau.[7] Tharsis is home to a dozen or so extremely large, but dead, volcanos, including Olympus Mons. It is likely that Tharsis was built up over long periods of time by fluid volcanic activity. Eruption after eruption, year after year, piled layer upon layer of fresh volcanic rock higher and higher to create the raised plateau we see today.

But why so much volcanism in just one place? On Earth, we get two kinds of volcanoes. The first kind are plate boundary volcanoes, like Mount Rainier near Seattle or Eyjafjallajökull in Iceland. Seattle is located next to a subduction zone where two tectonic plates are coming together and one of these, the Pacific plate, is descending beneath the North American plate. That descending plate brings water into the mantle. That water, in turn, changes the properties of the rock it encounters, allowing it to melt more easily and making it more buoyant. In Iceland, two continental plates are coming apart, which allows rock from the mantle to well up in between them, building a structure called the Mid-Atlantic Ridge. You can recognize this feature on globes that show the seafloor—it looks like the stitching of a baseball.

But volcanoes on Mars are of the second type; these are called hot spot volcanoes. On Earth we know that hot spot volcanoes like Mauna Loa, on the island of Hawai'i, are located over plumes of rising mantle material. This type of volcanism is less explosive and violent than plate-margin volcanism because there are fewer dissolved gases. The lava itself also tends to be much less viscous, which lends itself to forming volcanoes with a broad, shield-like shape rather than a pointed cone, like Rainier. As our tectonic plates slowly move over the hot spot, whole archipelagos and island chains are created.

But what if the crust didn't move? Something like this seems to have happened on Mars. Without any apparent plate tectonics, the hot spot–generated lava just keeps on coming and coming. Eventually, a huge volcanic province like Tharsis is produced. There can be something similar on the Earth if you have very fast erupting volcanoes. This type of eruption is sometimes called trap volcanism. In particular, the Deccan Traps in India are over 1,000 km across. This kind of volcanism is also very common on the Moon where you can see that lava filled some of the larger craters.

## PUTTING A NEW SPIN ON MARS

Tharsis is so big and so heavy that it likely affected Mars's spin. All planets, and indeed all spinning things, like to rotate with their heaviest parts located as far as possible from the center of the spin. This is called a "major axis rotation." You can test this at home by trying to spin different objects on a tabletop. Wider, more squat objects are more stable when they spin than are tall and skinny objects, which rapidly start to tumble over.

For planets, this means that if you build a huge volcanic province anywhere else but right on the equator, the planet will reorient itself over time. Because the whole planet moves along with the big heavy land mass, the north and south poles get pulled into new locations. This gives this reorientation process its name: true polar wander.

Not surprisingly, Tharsis is centered right on the equator. Something similar also appears to have happened on Pluto, where a large mass anomaly (a.k.a. a heavy flat area, like Tharsis) called Pluto's Heart lies directly on the equator. It's conceivable that both features formed right where they are, but more likely that they were originally formed in random locations and have moved over time. On Mars, other geological evidence suggests that Tharsis formed[8] at about 20°N (see figure 2.1).

## A WARM AND WATERY PAST?

While rappelling down the side of Valles Marineris, our daydreamer alludes to the possibility of liquid water on Mars in the past. Indeed, over

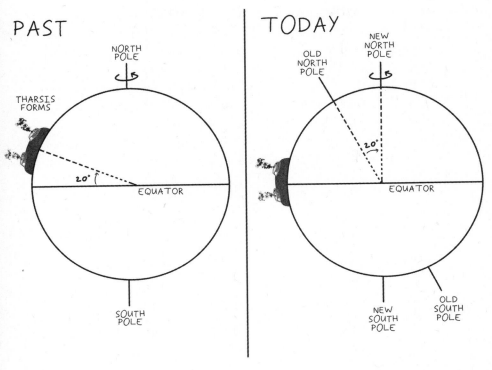

Figure 2.1
Evidence shows that the Tharsis volcanic plateau formed at 20°N latitude. Over time, the weight of the massive geological province caused Mars to reorient to a more stable configuration with Tharsis on the equator, changing Mars's spin pole in a process called true polar wander.

the past few decades, evidence has begun to pile up that indicates Mars may have been much warmer and wetter early in its history. So much so that many planetary scientists now believe that between approximately 4.1 and 3.5 billion years ago, Mars may have supported a global ocean of water, much like Earth.[9] While the details of this possibility are still being debated in the literature and at conferences, there are many intriguing lines of evidence. These range from the famous "Martian dichotomy" that may be a paleoshoreline to the very recent discovery of fossilized mud cracks in Gale Crater.[10]

It's exciting to imagine Mars with a global ocean. Life most likely started in the oceans on Earth, so why not on Mars? Unfortunately, we'll never be able to swim around and look for life, as the oceans dried up billions of years ago. Maybe we'll find fossilized evidence of past life that used to call these oceans home, and Mars will turn into a destination for planetary paleontologists. Or maybe these oceans never supported life.

While there is no global ocean on Mars today, it turns out that present-day global oceans of water are more common in our solar system than you may think. And this brings us to our next destination, Europa, a moon of Jupiter whose surface crust is made not of rock but of water ice. And hidden below that surface of ice is a global ocean of liquid water.

# BENEATH THE EUROPAN ICE

The sound of the icequake hits you like a thunderbolt, coming out of nowhere. You can sense the passage of the pressure wave through its sound, even if you don't feel the shaking directly. Like a siren speeding onward into the night, the pitch of the sound drops as it passes with high-pitched ricochets, reverberating all around you, reflected from every imperfection in the surrounding ice. It's a bit like the sound of a box of spilled spring coils washing over you. Just another reminder that, unlike some other planets and moons in our solar system, Europa is very much alive.

You're not getting back to sleep after that, so you sit up. The room lights, sensing your motion, ramp up their brightness. Momentarily, you are overcome by a sense of drowning as the light deeply illuminates the ice that makes up the walls and ceiling of your bedroom, submerging you in a vast oceanic blue. But, after a deep inhale, the feeling passes. Many of the other residents of this research station have painted their walls and ceilings to avoid this feeling, to give them a sense of safety in their private spaces.

But your preference is to take this moon at face value, to embrace the sublime and the strangeness in this place—even if, sometimes, it challenges your mental health in small ways.

Later that morning, you are working your way through the warren of the station. The rooms and corridors, more properly tunnels, were excavated out of solid water ice. It's an ideal system—excavation requires nothing more than a source of heat to liquify the native building material. In return, the human residents here get living space, samples to analyze, water to drink, air to breathe (once the ice is electrically separated into hydrogen and oxygen), and protection from the intense radiation of Jupiter's plasma torus.

The ions caught within that feature are deadly to humans and to spacecraft electronics. In the ancient days, the Europa Clipper mission was designed to dive close to Europa at high speed to collect ground-breaking data. Then, just as quickly, the spacecraft would escape to live, transmit, and probe another day. However, to mitigate all this radiation, all you need is a few meters of ice over your head.

But why build here at all? Europa has one of the largest research stations in the solar system because this place is special, with features not seen anywhere else. The surface ice of this world is cracked and crisscrossed with fault lines from tectonic activity. There is evidence of melt-through in places called "chaoses" where the surface looks to have been turned into a series of floating ice pans that became jumbled, like mixed-up puzzle pieces, before refreezing. Most tantalizingly of all, the Galileo spacecraft observed strange magnetic field properties, which suggested the presence of a salty ocean underneath the ice.

All this activity is being driven by Jupiter's intense gravity and the close resonant orbits of the four large Galilean moons: Io, Europa, Ganymede, and Calisto. Each moon passes the others at regular intervals and, as they do, they play gravitational tug-of-war with the giant in the center. This causes the moons to flex back and forth, deforming and relaxing their interiors in each orbit. Just like with a squash ball, that repeated flexing causes the moons to heat up from within. The results are spectacular. For instance, Io, closest to Jupiter and therefore experiencing the most intense squashing, is the most volcanically active body in the whole of the solar system! All its ice has boiled away.

On Europa and Ganymede, the next two moons out from Jupiter, the tidal energy translates into a liquid water ocean in the interior. That ocean is why you are here. Liquid water seems to be an important ingredient for any kind of life that we can imagine. Any place where it exists is, at least in theory, habitable. On Earth we have plenty of water to go around, whereas Venus and Mars have lost most of their water. But far more liquid water exists within the Galilean satellites combined than is in all the Earth's oceans. It doesn't take much imagination to realize that the Galilean moons represent the largest habitable volume in the solar system.

But just because a place is habitable doesn't mean that it is inhabited. Some think you need to have the sorts of interactions that occur on the surface of a solid planet to get the right chemistry for life. If that's the case, you won't find anything in the deep below. But you are an eternal optimist.

That's just as well because today will be your first dive. You are careful to control your nerves as you descend through a shaft deep into the ice. At first, the bare ice wall of the shaft is all you can see, but as the pressure outside increases, reinforcing materials from home secure your descent. The control room is above, in the safer part of the ice near the surface, and you check in with your support team as you descend deeper and deeper.

That pressure climb is inexorable. Europa's gravity may be less than one seventh of what you experience on Earth, but you still have a tremendous amount of ice over your head. Each meter that you descend adds another 1,200 Pa to the pressure trying to collapse the walls of this shaft. At that rate, you add another full atmosphere of pressure every 80 meters of descent. You're lucky that past spacecraft found a place on Europa where the ice thins out to just over three kilometers.

Even so, by the time you reach the bottom of the shaft, the external pressure has risen to what you would experience under 1,200 feet of water back on the Earth—a true abyss. Down here, the walls are metal, not icy, because they extend out beyond the base of the ice sheet and into Europa's ocean. You take a minute to gaze out of a clear porthole into that ocean—your first view of free water on the planet. But you are disappointed by the view—all you see is an inky blackness. What little light makes it out to Jupiter's orbit is wholly absorbed and extinguished by the three kilometers of ice above.

Turning on lights on the outside of the gantry corridor does not make the situation much better. This step is required for safety's sake. Just like on the underside of the Antarctic ice shelves of Earth, with so much water so close to its freezing point, any ice crystals that form will "snow" upward until they are lodged against the underside of the ice shelf. On a particularly "snowy" day, the whole launching tube and the submarines here can be completely encased in the fluffy white precipitation.

But today there is no need for any de-icing operations. You walk a short distance along the corridor and come to the floor lock for your submarine. The metal is well worn and shiny by many hands—you are not the first scientist to make use of the facility. But you are still excited by your first time out. If you close your eyes, you can imagine this gantry tube sticking out from the ice shelf and swaying in the Europan currents.

Punching up the display on the wall above the lock, you check the sub's systems. You are rewarded by indicators of full charge, full oxygen, and green nominal lights across the board. You report the conditions to the operators above, even though they too can see the same indicators that you can. Repetition, procedure, and multiple eyes all make for safe operations; here, in this distant place, rescue or resuscitation are unlikely in case of an accident.

As you open the lock, you get your first view of the sub's interior. In essence it is a sphere of thick, high-strength, clear plastic, which doubles as insulation and is augmented by a metallic coating along the sides to maximize heat retention. Instruments, sample collection equipment, and propulsion are strapped to the outside. Inside, there is a lightweight chair due to the low gravity, along with oxygen tanks and a computer system. The most heavy-duty component of the system appears to be the lock itself, sitting atop the submarine's plastic sphere and giving the whole submersible the look of a holiday tree ornament. Unlike on Earth, where such a sub would be boarded on the surface or in a moon pool, Europa's ice sheets mean that you need to enter the sub under pressure.

Looking down through the lock, you can see straight through to the ocean, the beams of light from the gantry cutting through the nighttime liquid until they are lost in the distance. Bits and pieces of ice or other particles sparkle as they enter the beam and then disappear once more. It's a vertiginous view, but you swallow back your hesitation and lower yourself into the seat, stowing a small satchel of snacks along with pencil and paper for the trip.

As you do so, you review your mission plan. The sub has considerable range, and today you'll be exploring a new sector on the floor below as part of the station's long-term plan for mapping. Some of this work can be done with robots, but the government consortium back home that sent you feels that there is value in having human eyes on the scene. You tend to agree.

Your check is complete, and with the permission of your local mission control, you initiate the launch sequence. The hatch above closes and shortly you receive a slight push that lets you know you are floating free of the gantry. You power up the props and head off into the unknown.

As the station recedes from your view, you take in the strangeness. At first you proceed along the underside of the ice sheet. Just like in your room, there are places with that deep glacial blue, ice that is startlingly smooth and translucent. In other places, the surface is pitted and cracked or scalloped on the underside by unknown processes. Periodically, you can see accumulations of gases emitted from far below. These gases pool against the ceiling, and the interplay of light refraction makes them appear like mirrors to you, reflecting the image of your sub.

Most sub pilots prefer to travel along the underside of the ice sheet to provide a fixed frame of reference before dropping down to the bottom. But you know that there have been enough sub journeys that the positioning systems can be trusted. You decide to drop into the mid-water and begin descending farther. Sonar scans of this area show that there is nothing for you to hit within kilometers, so there is no danger. At this point you have traveled so far from the gantry that you can no longer see any lights that are not your own. Those beams illuminate only the empty water. You wouldn't even know you were moving if it wasn't for the few particles that flow around you as you pass.

The journey will be a long one, so you spend your time alternately drawing mythic sea creatures on your pad and napping.

Eventually, you reach your diving coordinates. Though you were warned, nothing could have prepared you for the blazing white of the ocean floor. The pressure inside Europa is so high that exotic high-pressure ices with different crystal structures are occasionally formed. Unlike the kind of water ice that humans know and love, high-pressure ice sinks instead of floats.

If Europa's seafloor were completely flat, that would be the end of the story. Europa's ocean would be nothing more than a sterile watery world trapped between two layers of ice. You can't have life when your chemistry is limited to a single molecule: $H_2O$. However, just like any planet, encased in ice or not, Europa's rocky mantle has bumps and cracks, mountains, and volcanoes. In places, these features break through the high-pressure ice and provide oases of complexity. Here, salts and gases can dissolve into the water and heat is injected into the ocean from below.

This pattern also occurs back on Earth where the seabed is a barren wasteland except for the areas around volcanic vents. These vents are clogged with strange creatures that are totally alien to a surface dweller but are no less related to us than any other organism on the planet. Some scientists, arguing based on genetics, even suggest that these environments could have been the cradle of life on our own planet. Could something similar be happening here?

You ponder this question as you travel along the white floor of the ocean. You're headed for a volcano to see if you can find any unusual chemistry or shapes that might hint at life or its precursors. Previous exploration hasn't found anything, but the influence of luck in scientific fieldwork is not to be underestimated.

But before you get to the volcano, you are surprised to see a large crack that has opened in the ice below. That wasn't on the surveys.

As you approach, you can see that the water near the crack is shimmering—the heat-haze no less recognizable for being in water instead of air. This feature must have formed recently—yet more evidence that things change in this place.

You approach carefully: it wouldn't do to melt a hole in your sub. Relying on your external temperature probes, you find a cooler end of the crevasse to begin your inspection. Slowly, you descend into the feature below the level of the sea floor, taking measurements across all the wavelengths your instruments can sense and lidar scans of the surface topography. Once the initial reconnaissance and documentation is complete, you can indulge your curiosity.

Toward the center of the crevasse, you can see a new volcano being born. Strewn around the floor of the crevasse are smooth rocks of similar size—pillow lavas. The vent itself is a riot of different colors organized into a squat organ pipe shape, a testament to the variety of chemicals in the hot gases and how quickly they are cooling and solidifying into complex shapes. You can even see a faint glow.

On a whim, you decide to turn off your lights. You let your eyes adjust slowly and realize that the internal instruments still provide some illumination. You toss your jacket over the console and there it is—the faint reddish-orange glow from freshly deposited lavas. You smile, but as you pan away from the vent, your smile changes to surprise. The vent is not the only light you see. There is also a faint greenish glow surrounding the vent. That's not a volcanic effect; instead, it looks like bioluminescence!

Still, best to not get ahead of yourself. There are a surprising number of chemical processes that have nothing to do with life, but which build similar structures and have similar effects. Just try to build a definition of life that recognizes the nonliving nature of fire! No, collecting some samples will be the best way to proceed. Then your colleagues in the lab can decide whether this phenomenon deserves greater investigation or is simply a beautiful display of nature at play.

To better see what you are doing, you keep your lights off. You take a water sample and then try pulling a second sample through a fine filter just in case these glowing particles are suspended in the water. Satisfied, you decide to head back for home.

As you rise above the crevasse, you bring your lights back up to full power for the journey. But just as you reach down to remove your jacket from the instrument display, a sudden bright reflection of your headlamps on something from outside casts light throughout the sub. You feel a pressure wave jostle your craft and you look up in time to see something that looks like a dark shape dissolving into the gloom.

Were those tentacles?

Your heart skips a beat. The analytical part of your mind wonders: Could something be feeding on whatever is causing the bioluminescence? Something big? As you head for home you reflect that you have no idea what you just saw or whether it was really there. But here in Europa's hidden ocean there is room for the imagination to run wild.

★★

With the relative familiarity of the Moon and Mars behind us, we can take a larger step (swim?) out into the vastness of the solar system to experience something a little different: the icy moon Europa.[1]

### LAKES AND OCEANS IN THE ICE

Europa is one of the many moons of Jupiter and also one of the four *Galilean moons*, which are Io, Europa, Ganymede, and Calisto (listed in order of closest to farthest from Jupiter).[2] They are called the Galilean moons as they were the first natural satellites ever discovered in our solar system, and that discovery was made by none other than Galileo Galilei.[3] The whole Jovian system is about 5.2 au from the Sun, measured from the Sun to Jupiter. It wouldn't make sense to quote Europa's specific distance from the Sun for two reasons. First, Europa is 671,100 km from Jupiter, which is 0.0045 au. That's such a small difference from 5.2 au. Second, don't forget that Europa orbits Jupiter (once every 3.5 days), so it's constantly changing its distance from us. It's just easier to quote the Sun–Jupiter distance.[4]

Europa was one of the first in a category that is quickly growing in number called icy moons. Surprisingly, once you are far enough away from the Sun, water is relatively common; many natural satellites around the gas giants have water as a large fraction of their mass.

Of course, being so far from the Sun means it's cold, and you would expect that water to be in solid form (i.e., ice). However, over the last few decades, we have started to learn that this expectation is subverted time and time again. In turns out that vast oceans of liquid water likely lie below the surface of Europa and many other icy moons and thus represent some of the most exciting unexplored places in our solar system. Figure 3.1 shows a cutaway of how we expect Europa to be layered. It is structured with an icy outer shell, below which a liquid water ocean exists. This all sits on top of a rocky mantle and possibly some kind of metal-rich core.

Now, let's not get too excited because there are several issues with the logistics of exploring these oceans. As mentioned, the liquid water oceans exist *below* the surfaces of these icy moons. There are no exposed rivers,

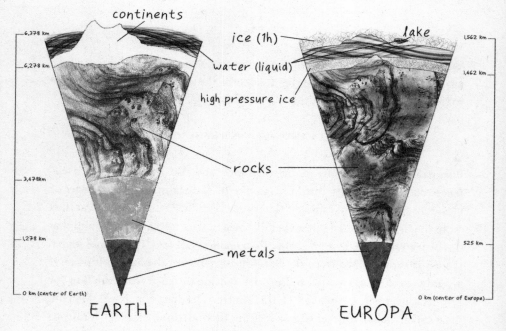

Figure 3.1
All planets and many moons are large enough that they have separated out into distinct layers made up of their individual components, though for some planets (notably, Saturn) this process continues to this day. Different planets are made from different stuff, however, as can be seen in this side-by-side comparison of the Earth and Europa.

lakes, or oceans. So not only have we not mapped the oceans directly, but we also only know they exist through indirect methods of investigation.

So what clues do we have as to the existence, size, location, and other important parameters of a liquid ocean beneath the surface of Europa? The most tantalizing pieces of evidence came from the Galileo spacecraft, which orbited Jupiter from 1995 to 2003. Aboard Galileo was a magnetometer, an instrument designed to measure the strength, direction, and changes in a magnetic field. This instrument made measurements as it passed by each of the large moons in orbit around Jupiter—Io, Europa, Ganymede, and Callisto. For Europa and Ganymede, those measurements were consistent with the presence of a salty water ocean.[5]

But wait, there's more! The geology on the surface of Europa imaged by the same probe detected signs that Europa's ice crust is thin. The surface is broken into moving tectonic plates with the same double-ridged structure we see on Earth at locations where plates are pulling apart, like the Mid-Atlantic Ridge. This suggests that the ice is thin enough to break and move around. We also see colors on the surface consistent with dredged-up salts. Most convincing, there are places where the ice has clearly melted through. These places are called chaoses for the random way in which rafts of ice have refrozen. It's clear they must have floated freely at one point to look like the scattered puzzle pieces we see today.[6]

It is possible to create the appearance of plate tectonics without the presence of a liquid ocean. For example, if there were no liquid ocean there, but instead a layer of solid ice all the way down until you get to rock, some of these phenomena could be produced with the convection of solid ice from deeper down toward the surface, just like Earth's mantle. However, this doesn't explain the presence of a magnetic field. Regardless, the planetary science community generally agrees that a liquid ocean is the most likely explanation for all these observations. More recently, water has even been seen geysering from between the cracks of ice at the surface of Europa, which suggests there may even be vents that lead all the way up from the ocean within.[7]

The big question on everyone's mind is: Just how thick is that solid outer shell of ice? It's not an idle question. If that layer is thin in places, say, a few hundred meters, then it might just be possible to explore the Europan ocean within our lifetimes. However, estimates put the thickness at several kilometers, or maybe as much as 100 km. You would need a serious space operation to get a robot through that much ice. Of course, for the purposes of our daydream, we assumed a thinner layer.

We might be able to get a better idea of just how thick the Europan ice is in the next few years. Once spacecraft like the planned Europa Clipper[8] get the chance to study the satellite in more detail, we ought to be able to tell which places have thick ice and where the ice is relatively thinner. It's possible that there might also be lakes within that ice—pockets of water within the frozen crust that will allow us to sample the Europan ocean without having to melt all the way through the icy layer to the ocean below.

## AN ENVIRONMENT POWERED BY THE TIDES

Why are there liquid water environments in these cold and distant worlds at all? At 5.2 au from the Sun, the surface temperature of Europa hovers between −225°C and −150°C. In these frosty conditions water behaves less like a skating rink and more like concrete. While temperatures tend to increase as you travel toward the center of every planet or moon, there is a secret ingredient for the moons of the outer solar system that creates enough heat to partially melt the outer layers of ice. That secret ingredient is gravity.

Remember in chapter 1 that we discussed how tidal forces both act to create the Earth's ocean tides and are the reason the Moon's spin was synchronized to its orbital period? Well, tidal forces are acting in the Jupiter system too, but in contrast to the Earth-Moon system, the Jovian system has many large moons, and all the gravitational fields of these objects have conspired together to create a remarkably interesting pattern.

The Galilean moons, through the gravitational influences of both Jupiter and each other, have been locked into a series of what planetary scientists call orbital resonances. That means they travel around Jupiter with orbital periods that are simple multiples of one another. For example, for every orbit that Ganymede makes, Europa makes two orbits, and Io makes four. In figure 3.2, we can see that Jupiter, Io, Europa, and Ganymede are all lined up nicely in (1), which would be a time that tidal forces would be acting strongly. By (2), Io has already gone around once, but Europa is only halfway around and Ganymede just a quarter of the way, but by (3) Io and Europa line up again. Due to these regular gravitational pushes and pulls, each of these moons is forced to deform and relax their interiors, creating friction, which generates heat.

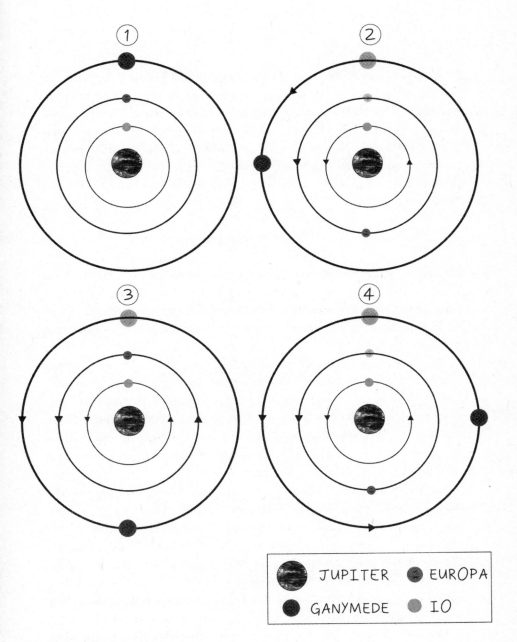

Figure 3.2
Over time, satellites orbiting close to planets tend to fall into regular patterns. Here
the orbital resonance of three of the four Galilean moons can be seen. For every orbit
of Ganymede, Europa completes two orbits and Io completes four. Note the original
positions of Ganymede and Europa are left as pale gray afterimages of themselves so
you can see how far each moon has gone since they started.

As we describe in the story, the effect of this "tidal flexing" heats up the interior of each moon. Io, being closest to Jupiter, experiences the most heating and has boiled away any water that was there to become the most volcanically active object in our solar system. We got a taste of that in our preface daydream. Next comes Europa, which maintains a thin shell of ice with liquid water underneath. Least affected is Ganymede, which has retained most of its water, leading to a thick ice shell and an even thicker liquid layer beneath.

ORIGINS OF LIFE

Of course, what's so exciting about these oceans is their capacity to host life. The salts on the surface of Europa suggest that even if there are high-pressure ices on the floor of the Europan ocean, there are places where the rocky interior is in contact with the water. That's important because life needs nutrients.

As you will see in chapter 10, all earthly life is primarily composed of just six elements, hydrogen, carbon, oxygen, nitrogen, phosphorus, and sulfur. These elements are arranged and rearranged into the basic building blocks, out of which every living thing is composed. The other ingredients that seem to be required are some kind of solvent like water (which allows things to mix easily) and a source of energy, both of which are abundant on Europa.

But let's not get ahead of ourselves. There's no evidence that life exists in the deep dark abyss of Europa. Moreover, we know very little about how life started on Earth (though we have many theories), but life seems to have begun almost as soon as the surface of the Earth became habitable. Did that life start in tide pools, powered by the sun? Or around deep-sea vents in the darkness? There's evidence for both theories.[9] But only the deep-sea theory would apply in a place like Europa. As a result, if we explore Europa and find life there, that result potentially tells us something about our own origins.

Let's switch gears, from a place that promises to host one of the most habitable environments in our solar system to a place that promises almost the exact opposite. Mercury is the closest planet to the Sun and is home to the largest daily temperature swing in the solar system. In such an extreme environment, let's imagine what kind of fun we could get up to.

# TREKKING MERCURY'S CREST OF DAWN TRAIL

Slowly, you awaken. Opening your eyes, you see a panoply of stars laid out overhead. Diamonds strewn across black velvet. At another time you could spend hours staring at this view in wonder, drawing imaginary lines between the points. But right now, it's something much closer than the jewelry-store case of the cosmos that grabs your attention: the thin shaft of sunlight illuminating the dust within your helmet. That sunlight has a single message: you slept in.

You sit up with a start and check your chronometer. It's an hour past dawn, well past the time when you should have started moving. To emphasize the point, you look to the east and see the wide curve of the Sun peeking over the horizon. The sight of the Sun so large is still shocking to the senses. It just looks wrong. From here, if the Sun was high in the sky it would appear to be 2.6 times wider than the Sun as viewed from the Earth with an area that could swallow up nearly seven full Moons. That might sound big, but you could still hide its disk with your thumb held at arm's length. Nevertheless, it's still unnerving, and you have yet to get used to such a sight, even two weeks into your hike.

Of course, the Sun has other ways of demonstrating its power in this place and you can already feel everything around you starting to heat up, though it is still early. Imperfect as it is, your suit's environmental system is keeping you largely comfortable right now. But that system can't easily fight the thermal load from a full and direct exposure to the Sun, not to mention the roasting oven of heat that would be emitted by every surface around you. At the peak of the day at the planet's equator, you know that the oven would be on the self-clean cycle, reaching temperatures of up to 800°F.

Luckily, the Sun rises slowly on Mercury. The planet and the Sun are locked in a strange dance called a spin-orbit resonance. It is like what we see with our own Moon on Earth, which rotates around its axis once for every orbit around the Earth. But because Mercury's orbit is much less circular than that of our Moon, the planet's dance with the Sun is much more complicated. Every time Mercury orbits the Sun twice, it spins around its own polar axis three times. The two motions combine to produce a day that lasts two Mercurian years, or 176 Earth days. That means your hour of delay has resulted in the Sun rising just 6 percent of its own diameter above the horizon.

Nevertheless, you have miles to go and it's time to start your day.

You rise to your feet and take a few minutes to stretch out your aching muscles. Often, those out trekking the Crest of Dawn Trail spend a few nights sleeping under the stars in their suits, as you did last night. This means that the suits trekkers like you wear are designed for flexibility as much as possible.

Once you are satisfied with your physical state, you take in some water through a feeding tube within your suit and some nutritious and energy-rich but texturally unpleasant substance meant to sustain you out here in between rest stops. You can't believe that you once thought runner's energy gel was bad!

Next, you hoist your pack, which is comically large and incredibly heavy—the pack easily out-masses you. But somehow the one third gravity of Mercury makes it manageable. That said, it's an awkward object and you need to take care with every step. A run or a stumble could rapidly lead to the pack continuing along under its own inertia and dragging you somewhere you didn't want to go.

That bulk is a necessary evil, however. Out here, you don't just need to pack your own food, water, and shelter. You need to bring your own air too.

If you were just lying around outside on Mercury, you wouldn't use all that much air—just 840 g of oxygen in a day, or about 3.8 kg once mixed in with an inert buffer gas like the nitrogen that we enjoy on the Earth. That may not sound like much, but decompress all that gas out to Earth sea-level pressure and you would have 3,000 liters of air. Trekking 60 km per day over the Mercurian landscape takes even more oxygen. In fact, it's not all that different from scuba diving. A typical scuba diver will get about 45 minutes below the waves out of a standard tank holding 6.5 lbs (2.95 kg) of compressed air. Do the math, and that works out to about 790 g of oxygen for every hour you exert yourself. For a full day, since you have a nitrogen recycler, that means the weight of about 10 kg of oxygen on your back, in addition to the tanks to hold that gas.

Of course, it's good to have some extra gas on hand—you can go for a few days without water, or weeks without food, but only minutes without oxygen. All told, you carry 40 kg of oxygen—enough to survive for up to three days of exercising out here, with a one-day reserve. On Earth, you'd need to bring 64 standardized scuba tanks to hold all that gas. Your nitrogen recycler drops that number to about 15 cylinders (14 oxygen plus 1 nitrogen)—still an absurdly high number, even in this gravity.

But luckily, you have another advantage out here on the surface of Mercury, walking in the darkness beyond the terminator, the line that separates night from day on a planet. The vacuum of space is very insulating, and for most of your trip the backpack is oriented toward interstellar space. Interstellar space, at about −270°C, is much colder than the boiling temperature of oxygen, −183°C, so your air stays pleasantly chilled. That means that instead of compressed gas, you bring cryogenic liquid, which packs much smaller. Now all your gas fits in the equivalent of just four scuba tanks. All you need is a little heater to bring the oxygen up to a breathable temperature as it enters your suit. Protect those tanks even further with a little bit of thermally reflective gold insulation, and you don't even need to take a compressor along to keep everything cold! It's like you are your own little spacecraft out here.

The reason for taking along only a three-day supply is because, though still an extreme wilderness, this particular trail on Mercury is a well-worn pathway. While relatively few people make this trek, those who do have stationed supply depots of liquid oxygen, food, and water spaced a few days apart. There is even a nice little shack at each supply depot where you can take the suit off and get a good night's rest.

That's where you are headed tonight. The idea of a warm shower and food that is not in paste form will give you extra motivation during today's hike.

Time to take a look at your map.

There are several trails on Mercury with different levels of difficulty. Each runs around the planet at a different latitude, like belts, meandering this way and that with the terrain to capture the sights and avoid the impasses. At 90°, the circuit would be ridiculously simple—a momentary spin around the north pole. But at the equator, the 15,300-km circumference pathway means that even a straight-line path would require a trekker

# MERCURY CREST OF DAWN TRAIL
## Along the 60°N parallel

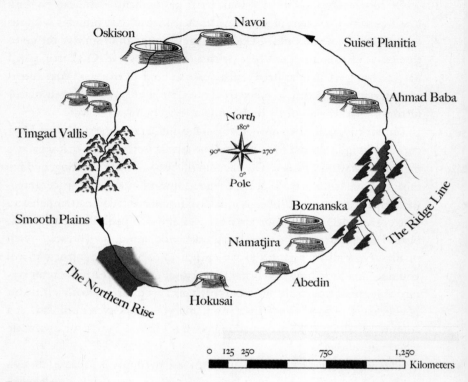

Figure 4.1
Trail map for the Crest of Dawn Trail, as viewed from above the north pole of Mercury (indicated by the cardinal rose in the center). The trail generally follows the 60°N line of latitude with deviations for terrain and scenic points of interest.

to hike 87 km per day to stay ahead of the Sun. In reality, given how broken the terrain is at this low latitude, you would need to average more than 130 km a day—a severe challenge for even the most experienced, elite athletes.

Instead, you are taking a moderate route. Here at 60°N you need to average only 43 km in a day along a straight line, a bit over 60 km along the trail. That pace would be elite level on Earth, requiring one to have literal wings like the mythological roman god Mercury. But here on the planet Mercury, with the reduced gravity, it is manageable by a mere mortal like yourself.

There are also different ways to complete the belt-line trails. All the trails are hiked east to west so that you travel in the same direction as the terminator travels. Despite the strange spin-orbit resonance of Mercury, this direction is the same as it is on the Earth. The easiest style of travel is a night-side passage in which you set off at Mercurian sunset and fall a little bit behind the terminator each day, finishing at sunrise, 264 days later. Those extra 88 days can come in handy, especially if you run into problems. Most Mercurian hikers start out this way—you yourself completed such a journey a few years ago. However, while there is an austere beauty to hiking out under the stars, this way of experiencing the landscape lacks a certain drama.

The next most challenging style is the crest of dusk, in which hikers chase the sunset and the dying of the light. While clocking in at just 176 days, the crest of dusk passage maintains the safety of the night-side passage. If ever you need a rest day, take a wrong turn, or find yourself slowing, you'll end up farther on the night-side of the planet. Indeed, at any point you can convert a crest of dusk passage into a night-side passage.

However, to call yourself a true Mercurian, you need to complete a Crest of Dawn passage. Here, the Sun chases you and you must use your knowledge, athleticism, and skill to pick your way across the landscape in fewer than 176 days. Should you fall behind for whatever reason, the Sun will climb in the sky until you are forced to seek shelter at a supply depot. There's little risk that you will be hurt—Mercury is, after all, the god of travelers—but you will end up stranded. In that case, your hike will come to an early and somewhat ignominious end as a rover comes to your rescue.

One of the joys of this kind of hiking is the freedom to choose your own specific route—connecting the dots between the different supply depots. And you must admit, you are pretty proud of the route you are blazing. Though it doesn't have quite as many highs or lows as an equatorial passage, and you don't pass through the well-known (but ultimately too big to appreciate from the surface) Caloris basin, the 60°N line actually has quite a bit of variety.

You've seen five different types of craters, walked on smooth volcanic plains, and have seen the infamous "low reflectance material" that shows up blue on color-enhanced maps. You've explored the mysterious "northern rise" and are looking forward to the views from the impact-formed ridge stretching from Namatjira to Ahmad Baba. Through it all you've picked your way among craters of all sizes and have skillfully found passes around or up past the contraction ridges that record Mercury's ancient crustal shrinkage.

Many weeks ago, your hike began at Oskison crater, right on the 60th parallel. Before setting off, you had already planned some time to explore the central uplift at the center of this feature. From the peak amid a chaos of hills you could see out to the crater rim, 60 km away in a beautiful 360-degree panorama. To the west lay a small hanging crater, suspended on Oskison's rim, pointing the way.

Onward you traveled, through valleys covered with impact-gardened volcanic rocks, softened, smoothed, and flattened by eons of broken rock from distant impacts raining down. Through Paestum Vallis and Timgad Vallis you wended your way along the floor of a labyrinth nearly 1,000 km long before finally emerging onto the volcanic plains.

At this latitude, the plains extend more than a quarter of the way around the planet, over 2,500 km. On this surface, your progress is quick, and so you get the opportunity to ramble to interesting spots. You deviate your course northward, to the south of the large crater Stieglitz and its heavily dissected ejecta mantle. You could get lost in there for days, so it's better to keep your distance. As you head onward, you feel the ground subtly rising beneath your feet, eventually topping out at 1,500 meters above the plains near the otherwise unremarkable crater Rivera. What caused this uplift on Mercury? Certainly, some planetary scientist knows the answer, but it's beyond your ken.

Past the northern rise, you head back down southwest toward the 60th parallel once more. Your destination is the crater Hokusai, an unusual formation not to be missed. As you approach, you find out why—the crater is incredibly fresh and there are mounds of large angular blocks and small secondary craters, formations that point back toward the impact point at the center of Hokusai in rays that extend for thousands of kilometers.

You know that Hokusai was once thought to be a shield volcano, so unusual was its appearance from the Earth. But here, you can enjoy the landform for what it is: angular blocks amid dark glassy impact melt like obsidian on the crater floor. Where Oskison had a jumbled central uplift, Hokusai has a ring of peaks near its center that form three quarters of a full circle. The gap in the ring invites you in, and you camp in the center for a few days to explore. You let the Sun catch up to you here, just to see the shadows cast by the ten or so individual peaks, like a conclave of stony giants, frozen in place on this unearthly plain.

Another 500 km or so of trekking takes you just south of a more conventional crater called Abedin. This crater has excavated material from deep below the surface, and you notice on the way from Hokusai that the plains here begin to change in color, darkening as more of the so-called low-reflectance materials are mixed into the regolith. But you can't tarry here—you gave up valuable ground to the Sun for that view at Hokusai and you have to make it up in advance of the most technically challenging section of the trek.

It has certainly been a memorable journey to this point, with sights you will recall for the rest of your life. The best, though, is yet to come: up ahead, the 60°N ridgeline beckons. A thin and meandering line, you intend to follow this high ground for nearly 1,900 km. From the lowest point on the smooth plains to the east of Namatjira crater, the ridgeline increases your altitude by nearly 4 km over less than 100 km, as steep an ascent as exists on Mercury.

With that, it's time to get up and get moving. On your last day here on the smooth plains, you want to put the Sun far behind you and to get to the depot near Namatjira as early as possible.

You reach the depot in the mid-afternoon. You recharge your oxygen tanks and load up on supplies, checking over each piece of your gear. You head inside to take off your suit and avail yourself of the shower. Feeling refreshed, you enjoy a luxurious meal made from the station's frozen supplies, making sure to log what you ate. While the water in the shower will be recycled, the rangers who maintain this station will need to replace the food that you have eaten.

Although you consider staying the night, you're too filled with excitement at the upcoming leg of the journey to stop just yet. In the eternal twilight here on the crest of dawn, humans revert to their internal circadian clocks and keep their own time. The length of our internal cycles varies, typically, from between 21 and 27 hours for different people. On Earth, the 24-hour cycle of daylight resets this rhythm each day, no matter whether you are a night owl or a morning lark. But here on Mercury, with no reference to time besides numbers on your chronometer, hikers go with what works best for their own biology.

As your circadian rhythm is on the longer side of things, there are miles you can travel before you sleep, and so you set off again.

Climbing out of the plains, the going becomes slower and the views become longer. While the ridgeline looks continuous on the map, with a

roughly constant elevation and a typical width of 30 km or so, its character
as the border between many crater rims means that the passage is filled with
benches, steps, and other blocky obstacles. That kind of enormous, inhu-
man detritus is to be expected in a landscape that looks like it was forged
and shaped by some god's hammer blows to the planet. At times, traversing
this terrain feels less like trekking and more like bouldering, if not mountain climbing. Over the next few hours, your problem-solving and route-
finding skills are put to the test.

The evenness of the elevation of the ridge could be just happenstance,
or it could be that the core of this feature is itself revealed as the crater rim
to an impact of truly titanic proportions somewhere far out on the boreal
plains. You suspect the latter, based on the subtle curvature of the structure.

For once you must manage the Sun, letting it rise enough to illuminate
the dramatic landscape, but not so much that it overheats your pack and
evaporates your oxygen.

As you pass north of Namatjira, you gaze into the central depression surrounded by a flat terrace—a very strange shape for a crater made even more
exceptional by the relatively bright volcanic materials it has excavated from
below the low-reflectance materials that make up the rest of the terrain.

Now, as you pass to the south of the larger crater Boznańska, you gaze
down over the side of a steep cliff falling over three kilometers to the floor
of the crater below. The famous hollows of this crater are illuminated
obliquely by the glancing light of the rising Sun, suggesting a meditation on
how they were formed. Was it something to do with the origin of the crater, with volatiles escaping the impact melt, leaving the hollows behind? Or
did the change come later, in some kind of volcanic cataclysm? Either way,
they stand out in this angular landscape shaped by the violence of impacts.

After hiking for the better part of 20 hours, you need to rest, and this is
as good a place as any. By letting the Sun rise a bit higher, you'll get even
better views in the morning. You find a small hollow and remove your
pack, being careful to shield it from the Sun.

Before you drift off, you reflect that over the past few days you passed
the midpoint of your trip. In a sense, it's all downhill from here. Ahead of
you lies the remainder of the ridge—the literal high point of your trip. But
in a few weeks, you will descend the spur that branches off southwest from
the ridgeline past Al-Akhtal. That spur returns to the smooth flatness of the
Suisei plains past the crater Ahmad Baba. On the rest of your trip, you will

see echoes of past sights, variations on a theme. Ahmad Baba has a peak-ring basin, much like Hokusai, though you know from the geology that it is similar in its materials to Abedin. In the same way as Namatjira presaged the rise into the heights, a similar crater in materials and structure called Navoi will signal that your journey is nearing its end back at Oskison.

But, with a smile, you realize that moment lies far in the future for you. Truly, it isn't about the destination so much as the journey itself. So long as you have a full pack of supplies, solid rock under your boots, a beautiful landscape to inspire and ponder, and a well-planned route to follow, you will be happy in the moment. That moment is all the sweeter with the Sun constantly chasing you, urging you along and reminding you that the moment is ephemeral. For that reason, you must savor each sight. Along the Crest of Dawn Trail, as in life, you can never retrace your steps, never go backward. Instead, the Sun allows you only to proceed ever forward, carrying with you all that you have learned from the journey thus far.

Here and now, as the Sun rises over the beautiful desolation of the Mercurian landscape, that is enough for you.

★★

Who walks around an entire planet? Such an ambitious hike sounds impossible, but you're thinking too Earthly. Here we have oceans that get in the way. But on Mercury there's no such obstacle. Mercury is the oft forgotten planet, and usually gets a shorter treatment in any space unit or undergraduate class than the other places in the solar system. Fortunately for us, our intrepid trekker circumnavigating a substantial portion of the Mercurian surface allows us an up-close and personal look at one of the solar system's most desolate and interesting locales.

### MERCURY, A GLOBAL PERSPECTIVE

Often, Mercury is visually compared to the Moon, and you can understand why: they both have a gray to dark-gray color, they are randomly covered in craters and mountainous regions, and they both have smoother regions, left over from past volcanic activity.[1]

Mercury is one of the four rocky—or terrestrial[2]—planets, all four of which occupy the inner part of our solar system. Starting from the Sun, the terrestrials are Mercury, Venus, Earth, and Mars (don't forget about figure 0.1, the solar system map). Mercury is the smallest of the rocky planets, and the smallest overall,[3] for that matter, at just 4,879 km in diameter.[4] It shares many similarities with its terrestrial family members, but it does stand out in a variety of ways as well. For example, unlike the other three, it has no atmosphere. To be fair, it does have an *exosphere*, a thin, very wispy shell of gas surrounding the planet made mainly of oxygen, sodium, hydrogen, helium, and potassium. This exosphere is 100 trillion times less dense than Earth's atmosphere at sea level and is constantly being blown away into a comet-like tail by the Sun's solar wind. Effectively, it is a vacuum at the surface.

But if the Sun is constantly blowing away this tiny amount of gas, why hasn't it been fully stripped at this point? That's because the Sun's radiation continually bakes off a small amount of the surface rock into gas to replenish that which is blown away.

On that note, the surface conditions are very harsh on Mercury. Daytime temperatures at the surface skyrocket to over 400°C. This is because, on average, Mercury is only 0.39 au from the Sun, which is 2.6 times closer than we are; so, it's bound to get hot! But without an atmosphere to regulate the temperature, the nights fall to a very chilly −180°C. Earth's Moon has a similar situation: a lack of atmosphere creates a large temperature swing between day (120°C) and night (−180°C). However, the Moon is hanging out with Earth, and thus much farther from the Sun. As a result, Mercury claims the title of the largest daily temperature swing in the solar system.

One of the biggest differences between Mercury and the rest of the rocky planets is the relative sizes of the internal layers. In chapter 2, we alluded to the layers of solid solar system bodies but have not yet fully defined them. Large objects in our solar system are usually *differentiated*. This means that as you travel from the surface to the center, you will encounter different layers made of different types of rocks/materials that act in different ways. Differentiation occurs because materials of different densities naturally separate out when they are allowed to flow and move. Less dense, or lighter, materials, such as rocks made of silicates (silicon and oxygen bonded together), move to the surface, and denser or heavier materials, such as iron and nickel, sink to the center. There are three basic components to a differentiated body: crust, mantle, and core.

All four terrestrial planets (and the Moon) have a silicate-based crust and mantle, with some sort of iron-rich core; however, Mercury's iron core makes up a much larger percentage of its volume than the iron cores of the other three. Inside Earth, the iron core has a diameter of about half the size of the Earth's diameter (there is a similar ratio for Venus and Mars). At Mercury, the iron core makes up over 80 percent of the diameter of the planet.

There are two major types of surfaces on Mercury: cratered plains and smooth plains. The cratered plains are the most abundant type of terrain and are characterized by stretches of rough, rolling plains pockmarked with all manner and sizes of craters. In between the craters, the surface is rough and rugged, and can have a broad range in high and low elevations. Most of the highest ground can be found in a band around the planet stretching to about 30°N and 30°S from the equator (like how Earth's tropic regions are a band straddling the equator). Of course, this is one of the reasons the trekker chose to walk a pathway following the 60°N latitude line, keeping safe from extraordinarily large changes in elevation.

The smooth plains are primarily found in the northern regions of Mercury and are comparable to the maria (large, dark, basaltic plains) found on the Moon. In both cases, these smooth plains were created by ancient volcanic activity, most likely kickstarted by large impacts that ruptured the crust and allowed lava to pour out onto the surface, creating the smooth impact basins we see today. The smooth plains are characterized by, well, smoothness, with very few craters, of course, but also the entire region is mostly about the same elevation. The exception to this is one of the more interesting places the trekker visits: the northern rise.

The northern rise is just as it sounds: within the relatively flat northern smooth plains there is an elevated region. Its height is about 1.5 km, but the change in elevation is subtly spread out over a very large area, about 1,000 km wide. You wouldn't really notice the rise was there while you were walking. So why is this interesting if you would barely notice it while climbing it? It's because we're not entirely sure how it was created.

One thing for certain is that the northern rise was likely created after the smooth plains themselves were created. We know this because the craters at the edge of the rise appear to have been tilted, as if they were already there before the rise occurred. Thus, the chronology suggests that large impactors and volcanic activity created the northern smooth plains; over time, craters began to pockmark the plains; and then some internal force pushed the northern rise 1.5 kilometers above the rest of the plains.

And that's why this is so interesting: What could have caused a massive 1,000 km-wide portion of the Mercury surface to rise up to 1.5 kilometers above the surrounding area? There are a few possibilities related to dynamics between the crust, mantle, and core of the planet, but the truth of the matter is still being debated by planetary scientists.[5]

## THE APPARENT SIZE OF THE SUN

To begin this story, our daydreamer helps us understand angular size on a much more visceral scale. When we look at any object on our sky, we see its apparent size, not its actual size. When something is nearby, it appears large, and when that same object is far away, it appears small. The Sun is an absolutely gigantic ball of plasma measuring about 1,400,000 km wide; that's 110 times wider than the Earth. But since the Sun is about 150,000,000 km away from us, it appears small. In fact, it appears to be about the same size as the Moon, which is only 380,000 km away (a fortuitous apparent size similarity that allows us to experience and enjoy total solar eclipses once in a while). So, from our vantage point on Earth, the Sun appears to be about the same size as the Moon, but if we were to get closer to the Sun—say, if we stood on Mercury, a distance about 2.6 times closer—then the Sun would appear larger.

But how do we know exactly how much the Sun's apparent size changes when viewed from different distances? The answer to that is pretty simple: it's just basic trigonometry. In figure 4.2, you can see the radius of the Sun and the distance from the Earth form a simple right triangle. Crunching the numbers, we get an apparent size of the Sun to be about 0.5 degrees.

This diagram also nicely describes why we use degrees as a measure of an apparent size or distance in our sky. To the human experience, the sky appears to be a giant dome on top of us extending from horizon to horizon. If you imagine the sky you see to be a perfect half sphere, then any object in your sky, or distance between two things in your sky, can be represented as some amount of angle.

Thus, we say that the Sun has an apparent size of 0.5 degrees from Earth's point of view. So, what would happen if we were on Mercury? With the same approach of figure 4.2, but using Mercury's distance of 57,909,000 km, we find that the Sun would appear to be 1.3 degrees in size. Thus, from Mercury's point of view, the Sun appears to be about 2.6 times larger

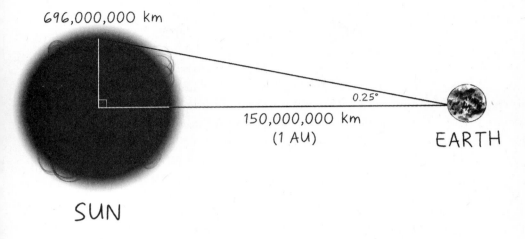

696,000,000 km

0.25°

150,000,000 km
(1 AU)

EARTH

SUN

Diagram not to scale

Figure 4.2
As you get closer to the Sun, the angle that our star subtends in the sky gets larger. On
Earth, this is about half a degree (0.25 degree for each half of the Sun). But on
Mercury, the Sun will be as large as 1.3 degrees across at Mercury's closest point in its
orbit. From this vantage point, the disk of the Sun would cover an area in the sky
equal to nearly seven full Moons.

than it appears from Earth. While this would be a noticeably larger Sun,
it would not be so big that you couldn't cover it with your thumb held at
arm's length.

You could also reverse engineer this approach and figure out how far away
from the Sun you would have to be for the Sun to appear to be any size.

### MAKING SENSE OF SPIN

The core plot device in this daydream is the trekker attempting to hike
westward around the planet Mercury, while keeping the Sun low enough
in the sky so as not to be turned into a human crisp. We called this trail the
"Crest of Dawn." As the Sun is rising in the east, the trekker must cover
enough ground toward the west to keep the Sun in the dawn position.

We can imagine what this would be like on Earth. If you were standing
on the east coast of Canada, and watching the Sun rise over the Atlantic
Ocean, you know that over on the west coast, they are still in darkness.

If you were to walk fast enough toward the west, from your perspective, the Sun would remain at the rising, or dawn position. Effectively, you must move as fast as, or faster than, the spin of the Earth to keep the Sun from getting too high in your sky. So how fast would you have to move to achieve this? Well, the Earth's circumference is about 40,000 km, and you want to move at a rate that matches the spin of the Earth in 24 hours. Thus, you need to travel at 40,000 km/24 h = 1,700 km/h! That is much faster than the speed of conventional passenger jets, and indeed even faster than the speed of sound.

So, if we wanted to do this calculation for Mercury, it sounds like all we need to know is the circumference of the planet and how long it takes for Mercury to spin once on its north-south axis.

Every body in our solar system spins around an axis, but not all spin at the same rate—or direction, for that matter. The gas giant planets have a pretty consistently high spin rate, with between 9 and 17 hours. Mars, surprisingly, has a spin rate very similar to Earth's at 24 hours 37 minutes. Venus, on the other hand, takes about 225 Earth days to spin once on its axis, and Mercury takes about 58 Earth days. These are both exceptionally long in comparison to the others.

So, we have Mercury's diameter (and therefore circumference), and we know how long it takes to spin once (58 Earth days), so we should be able to calculate the rate at which the trekker needs to walk. Right?

Wrong!

Interestingly, 58 Earth days is not the correct number to use here. That is because we have conflated *the time it takes for a planet to spin on its north-south axis* with *the time it takes for the Sun to do a full pattern of rise, set, rise again.* These are not the same thing! The former is called a *sidereal day* and is the time it takes for a planet to spin on its axis with respect to the background stars. The latter is called a solar day and is the time it takes for the Sun to rise, move across the sky, set, and then rise again (as viewed from a given location on the planet you are on).

For a planet that spins quickly, like Earth, the sidereal and solar days are almost the same.[6] However, Mercury's spin is so slow that the time required for Mercury to spin on its axis is an appreciable amount of the time it takes for Mercury to go all the way around the Sun, its year, which is 88 days long. Taken together, after completing one full spin on its axis, the Sun has not yet returned to the same spot in the sky. Figure 4.3 shows a depiction of

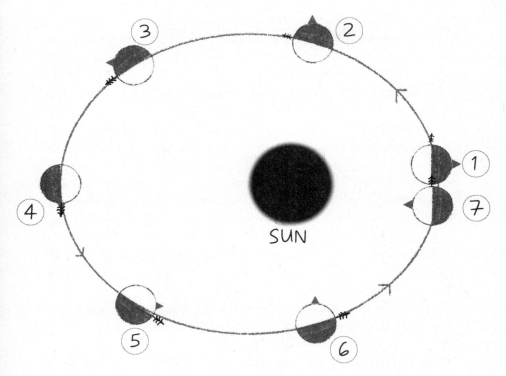

Figure 4.3
Just like our Moon, Mercury is locked in a resonance with the object it orbits, the Sun. However, because Mercury's orbit is elliptical (oval-shaped) the situation is more complicated, and Mercury rotates three times on its own axis for every two orbits around the Sun. In the figure, the center of the hemisphere facing away from the Sun at (1) is shown by the triangle and our trekker is shown by the tree. As the trekker follows the dawn terminator, the planet effectively turns underneath their feet. By the time Mercury has made one orbit at (7), half of the trail is complete.

Mercury's orbit around the Sun, with a small triangle representing a mountain located exactly facing away from the Sun (local midnight) in position 1, and a small tree representing the position of the trekker. How long does it take for the small triangle to be back to the same orientation, that is, how long does it take before it is pointing to the right on the page again? That occurs in position 5, about 58 days after position 1, which is one sidereal spin. But notice, that little triangle mountain has not yet come close to experiencing a solar day, which consists of sunrise, sunset, and back again to sunrise. In fact, after one full orbit of the Sun, when Mercury is back to position 1, the little triangle mountain has only finally gotten to the local noon position, pointing directly at the Sun. Projecting forward, that means it takes two full orbits of the Sun for any one place on Mercury to experience a full solar day, about 176 Earth days.

Getting back to our trekker, all of this leads to the question: How fast do they have to walk westward, to keep the Sun at dawn in the east (this is demonstrated as the little tree in figure 4.3 always remaining at the terminator)? If the length of the solar day is 176 Earth days, we have our time frame; now we just need to know the distance traveled. But we do have one last caveat: the trekker is not traveling the full circumference of Mercury, but following the 60°N latitude. This means they don't have to travel the full 15,000 km. If you maintain latitude at 60°N, traveling directly westward, it would be about half the distance all the way around. Thus, 7,500 km/176 Earth days = 43 km/day to maintain the sun's position. Is 43 km/day a possible distance for a human? There are at least two reasons why we know this can be done.

First, 43 km/day is roughly the same as a covering marathon of distance every single day, something that can definitely be done on Earth, as Canada's own Terry Fox so bravely demonstrated.[7]

Second, there are some comparable trekking distances on Earth. A famous example is known as the Great Western Loop, a walking path that loops around much of the western United States. It totals about 11,000 km. The record for the fastest completion of the Great Western Loop is 197 days and 11 hours, by hiker Nick Gagnon. Thus, Gagnon walked at a pace of about 11,000 km/197.5 days = 55 km per day.[8] That's faster than the pace needed to complete the Crest of Dawn Trail on Mercury.

But, in a lower-gravity environment like on Mercury, it should take less physical exertion to cover the same amount of ground on Earth. As humanity spreads through the solar system, maybe backcountry hikers will take up some of these interesting challenges. With a carefully chosen route, and lots of supplies, there's nothing stopping humans from hiking all the way around Mercury.

SPIN-ORBIT RESONANCE

Someone with a keen mathematical eye would notice that Mercury's sidereal spin, 58 Earth days, is exactly two-thirds of its orbit around the Sun, 88

Earth days. This means that for every two orbits around the Sun, Mercury spins on its own axis three times. Planetary scientists call this a spin-orbit resonance, and it does not happen by chance; it is the result of gravity.

In fact, we've seen a couple examples of spin-orbit resonances already. In chapter 3, we looked at how the Galilean moons are in orbital resonances with each other, but much more simply, in chapter 1, we investigated the Moon's spin-orbit resonance. In that case, we learned how the Earth's tidal forces have spun the Moon down to a resonance of one spin of its own axis for every one orbit around the Earth. Mercury is also feeling tidal forces, but from the Sun, which is also forcing it to spin a certain way, but it's not a simple one spin for every one orbit; it's three spins for every two orbits. Why isn't it a 1:1 resonance, like the Moon? This is because there is one significant difference between the Moon's orbit and Mercury's orbit: eccentricity.

Orbital *eccentricity* is the measure of how close to a perfect circle an orbit is, and is found through a calculation using the largest distance across the ellipse (a.k.a. the major axis) and the smallest distance across the ellipse (a.k.a. the minor axis). A perfectly circular orbit would have an eccentricity of zero. The closer the value of the eccentricity gets to 1, the more elliptical the orbit is. If the eccentricity is exactly equal to or larger than 1, then it is no longer an ellipse and we say the orbit is unbound (like a parabola, for example). No orbits in the solar system are perfect circles (thanks, Kepler![9]), but some are closer to circles than others. The Moon's orbit around the Earth has an eccentricity of 0.055, whereas Mercury's orbit around the Sun has an eccentricity of 0.206. In fact, if you look at a scaled drawing of Mercury's orbit, you might be able to see that it is elliptical. As a result of the shape of Mercury's orbit, which results in a large change in distance over the course of 88 days, the Sun's gravitational influence was unable to lock Mercury into a perfect 1:1 spin orbit resonance, but a nice 3:2 resonance was established.[10]

While the surface conditions on Mercury are certainly extreme, our next destination takes it to a whole new level. Even though Venus is much farther from the Sun than Mercury, it actually boasts an even more extreme environment. However, evidence is starting to suggest that Venus may not have always been so intense. Could it have maybe even been . . . Earth-like?

## PROSPECTING ON VENUS'S SURFACE

With a few taps of your rock hammer, the crystal comes loose. The sample is still quite hot, over 400°C, so you fit it into a specialized metal staff to take a better look. You plant the base of the pole into the ground and peer at the yellow light filtering from the sky through this transparent piece of halite-encrusted quartz, a solid crystal of salt wrapped around a silicate. There you have it—confirmation that Venus once had oceans just like the Earth.

Too bad you're a hundred years too late for that discovery! No matter—it's not pretty crystals that you're here to find. You're after a bigger discovery.

Still, in a small way, this crystal provides a clear window into the past. It tells you that this place is a good choice for hunting the fossils you're looking to find. Strictly speaking, looking for signs of past life on another planet makes you an astrobiologist. But you feel your intellectual lineage has more in common with the famed paleontologist Roy Chapman Andrews.

You turn your attention back to the crystal in its holder on the staff. Silently, you will the crystal to light up and shine a beam to the place you'll need to dig. But nothing happens. Maybe it's just that you're not actually Indiana Jones, the hero inspired by Andrews' exploits. Maybe you just don't have a clear sky to "activate" your crystal with a bit of movie magic. If that's what you need, you'll be waiting a long time. The forecast on Venus's surface is the same for the next billion years—hot and cloudy with a chance of sulfuric acid rain.

No matter, the search wouldn't be that interesting if a magic wand just showed you the way to go. You'll just have to figure it out for yourself. You release the crystal from your holder and toss it aside. The solar system moves on and last century's priceless sample is just another waste rock today.

Before giving up on this place, you trudge around in your suit, trying to get a better read on this site, high atop Fortuna Tessera, part of the Ishtar

Terra complex near Venus's north pole. It's not the easiest action for you to take. Your suit is exceptionally heavy, double-walled to protect you from the intense heat of this place. Its architecture has more in common with deep-sea diving than the typically skeletal structure of space equipment, as if you ordered the suit from a Jules Verne catalog.

You're happy to have it, though. You wouldn't last long outside under 93 times the pressure of Earth's atmosphere and temperatures that soar into the 400s of degrees Celsius depending on your altitude. Indeed, even electronics have trouble at these temperatures where there is nowhere outside your suit for waste heat to go. Unlike in other planetary environments where your time is limited by oxygen, here it's the special pack on your back that contains a wax-like substance that is slowly absorbing heat by changing from solid to liquid. Once the whole pack melts, your time will be up.

But you've still got a few hours until that happens, so you resolve to make the most of your time.

Moving around on Venus isn't just a question of hauling a bulky suit in gravity that's like what you would experience on the Earth. The combination of pressure and temperature results in an atmosphere composed of a fluid with a density that is nearly 100 times as thick as terrestrial air, or one-eighth as dense as water. It's not quite like swimming, but pushing your way from one place to another takes some work and doesn't happen quickly. For distance travel, you have a propellor pack with you that's not all that different from one used underwater; it's just made of the sorts of high-temperature alloys used in jet engines.

The increased density also plays with your vision. There may be few dust or sulfuric acid fog particles down here near the surface compared with the clouds above, but just like when you are underwater, the increased density scatters light efficiently so that faraway objects fade into the background more rapidly. The result is that, while the horizon is about as far away as it is on the Earth, you cannot see nearly that far. In the ocean back home, that visibility can be as high as 80 meters. Here, on a good day, at high elevation, you can get up to a kilometer or so.

That's a shame, because Venus has some of the tallest mountains in the solar system and you're close to them right now. The highest point of the range in Ishtar is Maxwell Montes, at more than 11 km above the plains below. Indeed, all of Ishtar is at altitude, and when you look at radar data from orbit, there are parts of this plateau that remind you of the Himalayas back home. On Earth, that kind of mountain range is formed by the titanic slow-motion collision of two plates of the crust. Perhaps a similar process did the same here.

However, unlike the Himalayas, which are a very young mountain range, these plateaus are ancient. Planetary scientists call them tessera, and they are the only rocks on the surface of Venus that record history older than 500 million years ago. At that time, a great cataclysm took place on Venus that covered the surface in lavas from the interior. "Crustal overturn," the geologists call it, and the impact craters that are evenly distributed across the surface are a testament to the similar age of all the planet's low-lying areas.

The tessera were spared because of their relatively high altitude, which results, in part, from their less dense rocks. In a sense, they float on the Venusian mantle like immense icebergs. It sounds ridiculous until you think about how the same process happens on Earth. At home, the continents really do float, but not on the waters of the ocean. Instead, the continents are made up of large complexes or "cratons" of less dense rocks, with a high content of a group of minerals called silicates. As such, they sit higher in the mantle just like the tessera, allowing the oceans to fill in the gaps between the continents. That water hides the denser, lower-lying, and much younger undersea basalts.

Similar rocks really do mean similar history, up to a point. To manufacture the tessera, Venus probably did have working plate tectonics much like the Earth and an ocean at one point. Given the size of the tessera, that ocean was probably around for a while. But, being closer to the Sun, at some point Venus entered a runaway greenhouse state. First, the oceans would have become more acidic, dissolving any limestone and other similar rocks, and driving carbon into the atmosphere. Then the oceans themselves would have evaporated entirely, which makes the process of recycling tectonic plates difficult. Like a jammed gear, deprived of grease, the pressure in the system built until it broke in a spasm: the crustal overturn that buried the lowlands and left only the tessera, standing high, to tell the tale of this vanished Venus.

Where are the oceans now? It's likely they escaped to space over time—we see chemicals in the atmosphere that suggest Venus lost a lot of water this way. This is similar to how the salt left behind once a pot of seawater boils dry can suggest how much liquid you had at the start. With the plate recycling broken, carbon dioxide released from volcanoes and carbonate rocks built up in the atmosphere, giving rise to the baked and dry landscape we see today on Venus.

It's a thrilling story. Maybe even a cautionary tale for our own home. But what interests you is what it means for life. If Venus once had oceans, plate tectonics, and cycles of mountain-building and erosion, it likely was a habitable place. If life is common in the universe when conditions allow, maybe ancient Venus wasn't just habitable but inhabited.

Mars might have started the same way, but probably became the arid world we see today rather quickly. Perhaps a microbial ecology could have evolved, but there may not have been time for much else. Venus, however, might have been a stable and habitable place for a very long time, longer than anywhere else besides Earth. Maybe even long enough for life to progress to something more macroscopic, something that could live and die and fall into sediments to become fossilized.

Something that a planetary paleontologist in a deep-sea diving suit like yourself could find!

Speaking of which, it's time to get moving. You turn on the electric motor to your propellor pack and rise a few tens of meters above the surface. It's a good way to survey a lot of territory in a short period of time.

As you start your surveying, you think it's strange how life imitates art. Hundreds of years ago, before the golden age of robotic planetary exploration, many thought Venus to be a rainforest. They could imagine anything was hiding beneath those thick clouds, and a verdant, lush landscape seemed the most likely with their limited knowledge of other planets. Fiction abounded with tales of imagined explorers having adventures in the Venusian jungle.

But the very first planetary probe discovered the surface of Venus was far too hot for any life we could conceive. Later probe landings confirmed it to be barren before orbiters arrived that could see in radar wavelengths. Those orbiters discovered the global cataclysm of 500 million years before, which put any evidence of what existed previous to that out of reach. And so, for decades, Venus was effectively abandoned as a target of exploration.

But then a renaissance and a return! Strange chemical imbalances observed in Venus's atmosphere gave some hope of finding life amid the clouds. Meanwhile, a better understanding of the surface geology started to sink in, and scientists realized that Earth's twin planet might have looked much more familiar until recently. Thus, Venus became one of the most interesting places to look for traces of past life. Who would have thought this planet would be such an important place for astrobiology?

You smile and think to yourself that the point at which a place or person is written off is often just the start of their becoming what they were always meant to be.

During this reverie, your surveying was more or less autonomous, but now you choose to focus your attention on the landscape. You pass over rocks folded like plasticine and beside coronae—low, circular hills where especially light materials like salts have risen inside the planet and cracked the supple surface. You look for the telltale signatures of sedimentary beds and decide to try something unexpected. Why not head up the side of Maxwell Montes in your search?

The idea is less silly than it sounds. The Burgess Shale, cradle of the Cambrian Explosion, is preserved high in the Canadian Rockies on a ridge of Mount Wapta. Even Mount Everest is full of fossils, former denizens of a shallow sea between India and Asia that was uplifted as the Himalayas rose. The remains of those organisms now find themselves resting at a higher altitude than any of their descendants.

Still, you're going to need all the luck you can find—this kind of prospecting and the people who do it are more than a little superstitious. In the hundreds of years since fossils were hypothesized on Venus, none have been found, and not for lack of trying. But maybe this place, Fortuna Tessera, is presciently named. Higher than all other tessera and largely ignored by other fossil hunters, it's a bit of a crazy place to go. But, you hope, a good crazy. Often, discoveries are made by those willing to explore the ignored places, searching against the common wisdom.

And so it is that you arrive at a jagged, unnamed outcrop, high on the slopes of Maxwell. You must be careful here: not only is the terrain rugged but everything is covered with a thin coating of iron pyrite snow—fool's gold. How appropriate, you think. Higher up, vast metal sulfide "glaciers" coat the mountain and prospecting is impossible. But here? Here you may have a ghost of a chance.

Carefully, you brush away the pyrite and take out a lamp to inspect the slope. To your trained eye, the surface is clearly sedimentary in origin, with fine layering. The surface is even a bit bumpy as you scrape lightly with a beryllium tool, as if something were embedded within the matrix of minerals below. It's a promising sign!

You unhook the sampler from your pack and set it against the rock face. It's the bulkiest piece of equipment you have but it is critical to your mission. A rock hammer would take hours here and you can't risk explosives. You press a few buttons on the sampler and the device whirrs to life. A few minutes later you pull it away from the rock face and it obliges, taking with it a perfectly polished hemisphere of rock. There's some fancy physics to how the thing operates, but that's not your expertise.

You look closely at the matching surfaces of the hemisphere and the hole in the rock, and smile. There is structure here—swirls and curlicues of different colors permeate the face of the sample like an arabesque. While it's possible there's an inorganic, chemical explanation for the patterns, you know life makes the same kinds of marks in the rock record when it's present. It's not an unambiguous sign, like a trilobite looking back at you, but you'll take it. Though you try to keep the growing excitement within you tamped down, you are not entirely successful. You remind yourself of the many times when others looked at shapes in rocks from other planets and, perhaps, saw more than was there.

"ALH84001," you repeat out loud to yourself, like a calming mantra.

Checking your monitors, you discover your heat sink is getting more melted than you'd like. There's just enough time to finish documenting the site and pack everything up. There's a spring in your step as you go about the well-worn ritual. Amid those checks, you take a bit of extra time and care to pack the sample more securely than is strictly needed.

There will be time to be excited later—in less than an hour you'll be back in your lab in one of the floating cloud stations here on Venus. You'll be able to shed this cumbersome suit. Then you can open a dialogue with your sample in the language of science. You can only hope that these rocks can share their stories. Perhaps you'll learn an as-yet-unknown chapter in the history of Venus. If you get that privilege, you'll relish the opportunity to pass that history along to your colleagues so that they too can have the chance to know this world as you do.

With that, you activate the propeller pack, and you are up, up, and away.

THE SURFACE OF VENUS IS A PRESSURE COOKER!

While it's possible for Mars to be closer to Earth than Venus at specific times during their orbits, on average, Venus is the nearest planet to Earth. Venus's average distance from the Sun is 0.72 au, and it is the closest match to Earth in physical size. Venus is 12,104 km in diameter,[1] where Earth is 12,742 km. As discussed in chapter 4, the rocky planets are made of similar materials, and Earth and Venus have similar layers, so this also means that the surface gravity on Venus is about the same (remember our discussion of surface gravity on the Moon?). Though, on the face of it at least, that's where the similarities stop.

For example, Venus's atmosphere is a wild place: it is both incredibly hot and incredibly dense in comparison to Earth.[2] The first direct measurement from the ground was done by the Venera 7 mission, launched by the Soviet Union in 1970.[3] It survived for only about 23 minutes due, in part, to the surface conditions, but one of the measurements it made was a temperature of 475°C, which is hotter than Mercury's highest temperatures!

Venus is closer to the Sun than we are, but not closer than Mercury. In fact, Venus is about two times farther from the Sun than Mercury. How is it possible that the surface of Venus is hotter than the planet that is closest to the Sun?

In chapter 4 we learned that without the presence of an atmosphere to regulate temperatures, Mercury's surface temperatures skyrocket during the day, but plummet at night. In contrast, not only does Venus have an atmosphere, but it is also very thick (much thicker than Earth's) and made almost entirely of carbon dioxide ($CO_2$), which is a gas that is really good at holding on to heat. Think of it like Venus is wearing a very thick coat that's really well insulated. On a cold winter's day, if you have a good enough coat with lots of layers, you could be sweating whether you're outside or inside. Venus's atmosphere is so good at holding on to heat that the temperatures don't even change that much from day to night. It's just hot on Venus . . . always. Mercury, on the other hand, with no atmosphere, has no such insulating layers. Any heat it experiences during its daytime is quickly lost to space during its nighttime.

On top of having a high temperature, Venus's atmosphere has a very high pressure as well. The first direct measurement of the pressure was done by Venera 8 in 1972, the next mission in the Soviet Venus program. It measured a pressure about 90 atmospheres (atm) or 90 times higher than Earth's sea level pressure.[4] This level of pressure is comparable to being one kilometer down in one of Earth's oceans. Meaning, at one kilometer under the water, if you were staying still, not moving, the push the water has on your skin from all directions would be the same strength as standing on the surface of Venus.

But this is difficult to picture; maybe there's another way to imagine this? Most humans don't experience pressures much different from regular atmospheric pressure at sea level, so it might be hard to imagine what walking around in an atmosphere with 90 times more pressure would feel like. You already know what it feels like to walk through Earth's atmosphere; if you go for a walk down the street you barely notice the atmosphere is there (if there's no wind). You might start to notice if you go for a jog: you can feel yourself running into the atmosphere as you run through it, resulting in a light breeze on your face. But if there were 90 times more pressure, things would be different. Indeed, the question confounded us a little when trying to find a simple way to make this more tangible. Let's start with some basics. What is pressure, anyway?

All atmospheres are made of matter, billions and billions of individual molecules zipping around at speeds somewhere in the range of 1,000–2,000 km/h. Those molecules move in all directions, even up, continually bouncing off each other, rocks, or anything else in their paths. What we experience as atmospheric pressure is a measure of the collective strength of the countless collisions of the atmosphere's molecules on a surface, say, your skin. In a planetary context, the atmospheric pressure is ultimately a consequence of the weight of the atmosphere itself. "Weight" is a term we use often in our everyday lives. It is the measure of how strongly gravity is pulling on some mass. As the gravity of Venus pulls all that gas down toward the surface, it squishes the gas into some amount of space, which determines how many molecules are bouncing around inside some given volume.

There are a variety of factors that govern atmosphere pressure, often represented together in the famous ideal gas law.[5] In figure 5.1 we can see some of the factors and how they act. Temperature ($T$), for example, determines how fast molecules move. Higher temperature means faster molecules, which equates to more energetic collisions. Thus, if you increase the temperature of a gas, you increase the pressure ($P$). Another important factor is the density ($\rho$) of the gas. The more molecules you pack into a volume, the more collisions there will be. Thus, if you increase the density, you increase the pressure. Venus has both a very high temperature and has packed lots of $CO_2$ molecules into its atmosphere. According to the ideal gas law, Venus

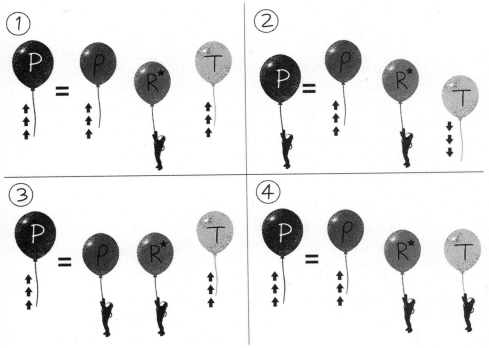

Figure 5.1
Gases at pressures that are familiar to us are governed by a useful relationship called the ideal gas law. This law states that the pressure of a gas is equal to its density multiplied by its specific gas constant ($R^*$) and the temperature. The illustration describes how the pressure ($P$) changes in response to changes in density ($\rho$) and temperature ($T$). For instance, as you approach a planetary surface, the gas gets squeezed like in a bike pump, which makes the density and the temperature rise. This in turn gives us higher pressure near the surface. In all cases, the specific gas constant stays the same.

must have a very high pressure at its surface, which is represented as case 1 in figure 5.1. In cases 2–4, we can see how playing with different variables in the ideal gas law can yield different results. For example, in case (2) if the pressure ($P$) is held constant (represented by an astronaut holding the balloon), and you increase the density of the gas, the temperature must go down. How do the variables play off each other in cases 3 and 4?[6]

So, using the equation in figure 5.1, is it possible to calculate how it would *feel* to be walking through an atmosphere that has 90 times more pressure than Earth? For starters, we can use the ideal gas law to calculate the density of Venus's atmosphere. Since we know $P$ and $T$ from two different Venera missions, we can rearrange the equation for $\rho$. This turns out to be 70 kg/m$^3$. Earth's atmosphere has a density of 1.16 kg/m$^3$, so Venus's atmosphere would clearly feel thicker than that (about 60 times thicker!). Liquid water, on the other hand, has a density of 1,000 kg/m$^3$, so it wouldn't feel like it was as thick as water when you go swimming.

Now that we know the density of the atmosphere, we can do one more calculation to work out what Venus's atmosphere would actually feel like to walk though. A good way to determine the experience of this atmosphere is to think of it in terms of *drag*, or in other words, the amount the atmosphere pushes back on you as you move through it. An average walking speed on Earth is about 1 m/s, and clearly, that does not generate much, if any, drag. But what if we were walking on Venus at 1 m/s? How much drag would that create?

This can be calculated using the drag equation.[7] The amount the atmosphere resists the movement of an object depends on the density of the atmosphere, the speed at which the object is moving, and the shape of the object. Crunching some numbers, we find that walking at 1 m/s on Venus would create the same amount of drag as if you were walking on Earth at 8 m/s. Now, 8 m/s is actually 30 km/h; humans cannot walk that fast! But try to picture the amount of drag you feel when on a bicycle at a good clip. There is a noticeable force of atmosphere against you as you pedal. So, walking around on Venus looking for fossils would come with a little more drag than you're used to while walking around on Earth.

FOOL'S SNOW

As you will learn in chapter 8, Venus has constant global coverage of clouds that are impossible to see through (at least with visible light). The only way to get a conventional picture of the surface is to get below the clouds, something that is very difficult to do in such an intense atmosphere. Luckily, there are other ways to "see" the surface, such as radar. Using radio light we can map the entire surface![8] We'll look at that more in depth in chapter 8, but we mention it now because these radio maps of Venus led to a surprising discovery.

The key here is that radio light will bounce off different materials in different ways. Some types of rocks are good at reflecting radio light, while others are not. So not only can radio light help you create a map of places hidden behind clouds, but by studying the brightness of the reflection you can also get information about the type of rocks the radio light is bouncing off and how they are arranged (their roughness).

Radar surveys of Venus have consistently shown that at the highest elevations, like on the mountains atop Ishtar and Aphrodite Terra, the radio light was bouncing back brighter than in other places. This leads to the question: How has the surface rock composition changed at these higher altitudes such that more radio light is bouncing back? Evidence seems to suggest frost, but not the kind you're thinking of.

Back on Earth, sometimes the temperature and pressure demand that $H_2O$ not condense out as liquid, but it must condense out as a solid. When that happens, we get tiny pieces of solid water falling from the sky, in the form of little crystals we call snowflakes. Solids can condense out of the Venusian atmosphere as well; however, the atmospheric conditions allow this only for specific chemicals at specific altitudes. Studies show that the temperatures and pressures in the Venusian atmosphere between about 2.5 and 45 km off the ground favor the condensation of molecules with various metals such as iron, lead, and tellurium. As shown in figure 5.2, these metals likely were baked from rocks at lower altitudes, where it's much hotter. They were then mixed in with the atmosphere where they can bond with chemicals like sulfur to form pyrite, and then can condense as frost when the conditions are right. The only place that can happen is at higher altitudes where both the temperature and pressure are lower. So as our prospector climbs the Maxwell Montes, it would make sense that they would start encountering pyrite frost at higher altitude. Maybe it even occasionally flurries in fool's gold![9]

### CURRENT HELLSCAPE, FORMER OASIS?

You might wonder why, of all the places in the solar system, we chose Venus as a place to daydream about searching for evidence of life. Certainly, with current temperatures and pressures, it is very unlikely there is life there today. But one of the more surprising revelations of the last 50 years of Venusian research is that in the past, perhaps even as recent as 1 billion years ago, Venus may have been cooler, wetter, and maybe even a bit Earth-like. Why do we think this? There are multiple pieces of evidence that lead us in this direction.

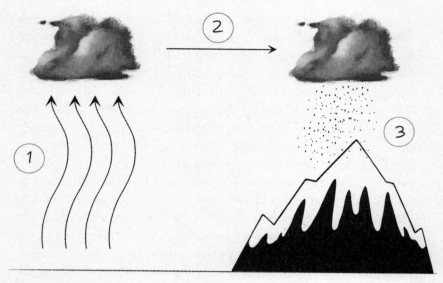

Figure 5.2
The high temperatures on Venus can evaporate metals and sulfur from the rock of the crust. Once liberated as gases (1), they rise and cool in the atmosphere, eventually condensing to form clouds (2) that snow on high mountain peaks (3). An identical process occurs on Earth but with water vapor in place of metal vapor.

First, in the early solar system while the planets were forming, the Sun was not as hot as it is today. This means that conditions on Venus would have been much more temperate. In fact, evidence suggests that at this early time, Earth may have been a bit too far from the Sun, making global oceans freeze over in what is called a "snowball Earth" event. But over time, the Sun has slowly gotten hotter, pushing what we call the habitable zone farther away.

Also, as the solar system was forming, Earth and Venus were not that far apart from each other, and in the grand scheme of the solar system, they should be made of roughly the same materials, including the amount of water present.

Second, radar maps have shown us that the surface of Venus kind of looks like Earth with all the oceans dried up. There are two very large, plateaued areas that one could interpret as ancient continents. The one in the north is called Ishtar Terra, and the one straddling the equator is called Aphrodite Terra. These two "continents" stand above the rest of Venus's

$$SiO_2$$

Figure 5.3
Over time, volcanism associated with plate tectonics can create abundant rocks rich in silicon dioxide ($SiO_2$). Called silicates, these rocks are less dense than typical volcanic rocks called basalts. On the Earth, silicates are concentrated in chunks of continental crust that float on the mantle like icebergs, while basalts are found on the bottom of the oceans. Venus has several high-standing regions that could also be composed of similar rocks, of which quartz is the purest form.

lowland areas, which one could interpret as the floors of dried-up oceans.[10] This is a nice idea, but is it true? Early on in our daydream, the prospector is walking around on Ishtar Terra and uncovers a halite-quartz mixture, something they cast aside as yesterday's news (see figure 5.3). But finding such a rock on Venus could confirm what we are beginning to suspect.

Venus had active volcanism like us (and still does!), and probably plate tectonics as well. These kinds of geological processes are made easier in the presence of water, which acts as a lubricant. Volcanic rock that has a larger amount of silicates in it becomes less dense. Thus, due to the presence of water, Venus's volcanos would create regular basaltic rock, which is more dense, and silicate-rich granite-type rocks, which are less dense. Rocks of two different densities will naturally settle, so the denser rocks sink down while the less dense rocks float up. Indeed, one can think of Earth's less dense continents of granite floating on top of the denser basaltic rock below. So, the very presence of continents on Venus suggests there were large amounts of water present being infused into rocks.

Third, the characteristics of the Venusian atmosphere also suggest it was a former ocean world. How do you get an atmosphere that is incredibly thick and made almost entirely of carbon dioxide? Planetary scientists think that a likely way to do this is through a runaway greenhouse effect. Let's imagine Venus started out with global oceans of liquid water and a thinner atmosphere, maybe even made of nitrogen, much like Earth. Studies show that even a modest amount of heating from the Sun could start to increase evaporation from the global oceans, letting $H_2O$ and $CO_2$ into the atmosphere. This would create a positive feedback loop, a runaway greenhouse effect, whereby the hotter it gets, the more evaporation occurs, putting more heat-trapping gases in the atmosphere, leading to even more heating. Eventually, you would evaporate/boil away your oceans, and even start baking the $CO_2$ out of the rocks, leading to an atmosphere entirely composed of carbon dioxide.

Some of these gases could escape Venus entirely. At the top of the atmosphere, a molecule can get so much energy that it flies away from its home planet forever. Other molecules can be split apart by ultraviolet and other high-energy radiation from the Sun with the smaller pieces flying off. Carbon and oxygen are relatively heavy atoms, and molecule fragments that contain either are hard to get rid of. However, hydrogen is light and can be very easily lost if it becomes separated from the other atoms in $H_2O$, so eventually it's as if the water was never even there. We know this happened on Venus because its upper atmosphere is full of deuterium (D)—the rare form of hydrogen occurring in all water that makes heavy water (HDO) heavy. The amount of deuterium we see makes sense only if Venus lost about as much water to space as the Earth has now in its oceans.

Finally, once water has boiled away, you lose the lubricant that makes so many geological processes easier. Any plate tectonics would effectively stop, jamming the gears, so to speak. But the planet's internal heat would continue to build inside. Eventually, if this is allowed to go on long enough, it's possible for the entire crust of the planet to reach a tipping

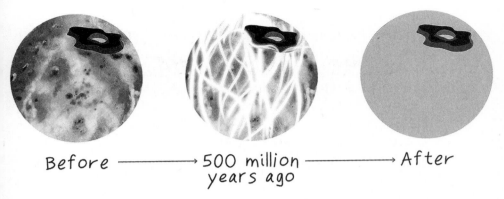

Before ———————→ 500 million ———————→ After
                      years ago

Figure 5.4
Ages calculated by examining craters on Venus's surface shows that most of the
geography is only about 500 million years old. This suggests that crustal overturn
may have taken place, in which the entire surface was replaced in a tremendous
episode of volcanism and tectonic plate recycling. The only areas spared from the
violence were the elevated highlands, which were able to preserve rocks from before
the overturn period.

point, whereby it cannot hold the heat anymore and magma spills out onto
the surface everywhere, all at once. This would resurface the entire planet,
save any places that were tall enough to stay out of danger. This is crustal
overturn. See figure 5.4.

## EXTRAORDINARY CLAIMS REQUIRE EXTRAORDINARY EVIDENCE

The idea of a planetary paleontologist, we admit, sounds like the coolest
job for young scientists to aspire to—not only exploring an alien world
but searching for signs of past alien life. Of course, this is either directly or
even indirectly one of the goals of many missions out into the solar system.
The Perseverance rover, for example, was sent to Mars with the expressed
purpose of searching for ancient signs of life. Is there life in the solar system? *Was* there life in the solar system? What would this mean to us? These
are big questions, and will redefine our view of ourselves, whether we find
evidence or not.

    With a question of such magnitude, it is important to be careful when
making claims. If a scientist genuinely wants to say they have discovered
life in the solar system beyond Earth, they should have good, if not irrefutable, evidence.

At the end of our daydream on Venus, the mantra our plucky planetary paleontologist repeats to themselves is "ALH84001." This speaks to an oft quoted phrase: "extraordinary claims require extraordinary evidence." When Carl Sagan said this, he was referring to the idea that claiming to have found evidence of life is incredibly important and requires proof equal to the level of the claim.

There have been a few notable examples where scientists debated publicly whether they had found life beyond Earth. Perhaps the most famous example was ALH84001, which is a meteorite discovered at Allan Hills (ALH) in Antarctica in 1984.[11] Any meteorite is a rock or metal chunk that was once floating through space, but then crashed into Earth. Planetary scientists love searching for and studying meteorites because they are from other worlds. It's like the universe has sent us a little sample straight to our door, without the need for building an expensive spaceship or endangering human lives to go get one.

ALH84001 is a unique meteorite because it is Martian. This means that this rock was once on Mars, as part of the ground. What likely happened is some large impactor smashed into Mars, ejecting some rocks straight out into space. ALH84001 was one of these rocks. It escaped Mars's gravitational pull and circled through the solar system for, perhaps, millions of years, until one lucky day it landed on Earth, and some scientists picked it up.

But when those scientists cracked it open and started looking at it closely, there were some microscopic features (smaller than 1 mm) that somewhat resembled fossilized bacteria. Some scientists took these features in the rock to mean that bacteria once existed on Mars, and were fossilized many millions or billions of years ago. The news of this discovery even reached the then President of the United States, Bill Clinton, who referenced the work in one of his speeches.

But if something looks like bacteria, is that enough evidence to claim that you have definitely found bacteria? The consensus among the planetary scientist community is that it is not. Many things can look like bacteria, and while the meteorite is intriguing, more is needed to confirm whether this is evidence of life.

Another famous example of some potentially intriguing evidence suggesting life beyond Earth is that of the Labeled Release experiment on the Viking missions to Mars in the mid-1970s.[12] The Viking 1 and Viking 2 were twin landers sent to Mars to, among other things, search for any possibility of life there. One of the ways they did this was to scoop up some of the Martian soil, add nutrients to it, and search for any possible signs of metabolization of those nutrients. If bacteria existed in the Martian soil, certainly they would be happy to receive some water and other yummy nutrients!

Incredibly, a really interesting signal was detected coming from the nutrient-infused Martian regolith: gases were given off that could be interpreted as arising from some type of biology. However, in this case, scientists were quick to point out that the type of gases that were coming from the soil could be explained by other methods. This means that, while the results were intriguing, life didn't have to be present to explain the results.

Relatively recently, in 2020, planetary scientists studying the atmosphere of Venus also made a really exciting discovery: they detected the presence of phosphine at high altitudes in the Venusian atmosphere (about 60 km above the surface).[13] Phosphine is a molecule composed of one phosphorus atom and three hydrogen atoms ($PH_3$).

On Earth, phosphine is both used for and a by-product of a variety of industrial practices. It also is created in a variety of natural ways: interaction of sunlight with certain gases, lightning, and also through some biological processes. For example, phosphine is a gas that is emitted by some bacteria when they metabolize.

When phosphine was discovered in the atmosphere of Venus, scientists quickly tried to account for all the possible ways it could be made on Venus. Clearly, there are no industrial practices there, but they investigated sunlight, lightning, and other possible pathways. Their result: there are no abiotic processes that can create the amount of phosphine they observed. Thus, it could be due to biological processes. They were careful not to say they found evidence of life, however. And it's a good thing, because shortly thereafter they realized they had overestimated the amount of phosphine in the atmosphere.

Currently, the exact amount of phosphine present in the atmosphere of Venus is unknown. What remains to be seen is whether the abiological processes present at Venus can account for any amount of phosphine.

All the above examples are intriguing, no doubt. They are interesting pieces of the puzzle. But taken on their own, or even all together, they do not present an irrefutable claim that life is present in the solar system beyond Earth.

Such a claim would be extraordinary! We must require its evidence to be equally extraordinary.

So far, we've looked at the Moon, Mars, Europa, Mercury, and Venus. All these locations have been moons or planets. But there are other types of places to visit, for example, planetary rings. What would it be like to fly through the rings of Saturn? Let's find out.

# SURFING SATURN'S RINGS

There is no greater teacher than firsthand experience. Therefore, it's not surprising that Saturn's rings are where the best pilots in the solar system come to learn their craft. Those rings are not solid surfaces but are instead made up of an uncountable number of individual particles. Some are no larger than a particle of smoke, while others are as big as a house. There are even bigger ones, hundreds of meters in size, called "moonlets" embedded in the snowy maelstrom. Meanwhile, many of the true moons of Saturn live within the gaps in the rings.

But a great many of the ring particles are human scaled, from snowballs to small boulders. That familiar scale allows an observer to discern up close the dynamical dance of all the particles moving together as they perform their timeless circuit around Saturn.

Today, you will be joining that dance and you can't wait.

Your classroom training has been building to this day. You have endured years of study and have logged thousands of hours in simulators back on Earth. You practiced with real spacecraft in empty space on the way out to Saturn. And, several weeks ago, you arrived on Pandora, a moon located just outside Saturn's F ring for what is, in effect, your final examination.

But first, like any good student, let's bring in the appropriate context. Saturn's rings are subdivided into different letters based on the order in which they were discovered. Giovanni Cassini was the first to observe a subdivision between the rings and named the outer ring the "A" ring and the inner ring the "B" ring. These are the two brightest rings, the ones you can see in a telescope image, and they are divided by 4,800-km gap that carries Cassini's name to this day.

Later observers found more rings: a dimmer structure closer to Saturn became the C ring; an even dimmer band, the D ring. Meanwhile, another band located farther from Saturn became the E ring. The first spacecraft to visit Saturn discovered a ring in the gap between the A and E rings, which, confoundingly, became the F ring—which means that from Saturn out-ward the lettering is DCBAFE. But what's in a name? In reality, all these different rings are a collection of many thinner rings, made up of even thin-ner rings, and so on, like the ridges on an ancient vinyl record.

All this structure is maintained by the shepherd moons, which orbit within the gaps and keep the smaller particles that exist in each ring on course with gravitational tugs. While Pandora, where your classroom is located, is not such a shepherd, the next moon inward, called Prometheus, plays this role. Indeed, the F ring is so young that Prometheus still collides with the ring at a particular spot in each orbit, like a dog nipping at the heels of errant sheep to keep the flock in line. The F ring shows the scars of these collisions, with dark lines and streaks where ring particles were drawn away by the encounter.

Because of this unsettled evolution in the F ring, you've been dropped off a little closer to Saturn at a quiescent spot nearly 137,000 km from the cloud tops of Saturn and about 100 km from the inner edge of the Roche Division, which separates the F ring and the A ring. From here you will chart a course that will take you in toward the moon Pan, circling within the Encke Gap. On the way, you'll pass by Daphnis, the shepherd moon that maintains the Keeler Gap. You've also been asked to find one of many interesting structures within the rings called a propeller, and to document it as if you were on a reconnaissance mission.

Just like how a sailor trains on a small dinghy instead of learning directly on a schooner or a cruise ship, you have been provided with only a very simple spacecraft. In fact, the spacecraft is so small that even you cannot fit within it and will therefore be in a spacesuit for the entire flight. This "pro-pulsion and support unit," as your spacecraft is called, is flat and oblong in shape. It reminds you of a comically oversized surfboard—which is what all the pilot candidates call it.

The surfboard has electric propulsion rockets, better known as ion engines, that use fuel very efficiently, throwing xenon atoms out the back at 40 km/s, faster than the Earth orbits the Sun. There's also power and consumables onboard to keep you comfortable. Luckily, you can clip in and don't need to stand and balance on the thing while it is under power. But on the positive side, you have unparalleled visibility for navigation and the flexibility of leaving the board and exploring just by using your suit, if the mood strikes you.

Just like with any spacecraft, there are pre-flight checks to complete, and then it's time to head off. A typical spacecraft would have a system in which a computer calculates the thrust pattern to take you where you want to go. But here on the surfboard every control is manual. It makes sense—the whole point is for candidates like you to get a feel for the orbital mechanics directly. Learning that celestial dance is not intuitive. With everything in motion, you can't just point yourself in the direction you want to go and expect to arrive at your destination. Instead, like meeting someone on a crowded dance floor, you need to take an indirect route.

You close your eyes and remember your training. There are four simple rules to moving around in an orbit. First, if you don't do anything and leave the engines off, you'll continue in the same orbit on which you started. Pretty boring, but good to remember since it means you don't have to spend fuel to keep traveling around a planet. Second, making yourself speed up in your orbit makes your orbit bigger. So, if you want to travel out away from the rings you should point your engines opposite the direction you are moving. Third, slowing yourself down makes your orbit smaller. So, if you want to travel inward, you'll need to point your engines in the direction you are moving to lose speed.

The fourth rule is that the place where you last used your engines will always be part of your new orbit. That means that if you want to add speed and travel back out to Pandora, the furthest you will be from Saturn is on the opposite side of the orbit, after which you'll fall back to your current position again. Perplexingly, even though you're using your engines to speed up, once you reach all the way out to Pandora on the opposite side of your orbit, you'd be moving even slower than you are now. It seems strange, but your instructors taught you to think of it like tossing a ball into the air and catching it again. The ball moves the slowest when it is furthest away from the Earth at the top of its arc and fastest when you catch it again. Still, it's a bit of a mindbender.

Well, since you want to go toward Pan, you need to slow down. So, instead of pointing the surfboard toward Pan or inward toward Saturn, you rotate yourself until you are edge-on to the rings and fire the engines to kill some of your forward momentum. A few minutes later, you smile. That must have done the trick because you start to see the rings approaching you.

From where you started, the rings were a relatively thin line. Even just 100 km away, the rings are just three hundredths of a degree in thickness, less than a tenth of the width of the full Moon as seen from Earth. Meanwhile, Saturn occupies fully one eighth of the entire circle of the sky, looming overhead with a width of over 90 full Earth Moons stacked on top of one another. Despite being incredibly broad at more than 270,000 km across, the rings are vanishingly thin, at only 30 meters or so thick on average.

They are also very bright. The ring particles are nearly pure water ice with very little contaminant material. It's the sort of structure that could only form from something like a moon being taken apart layer by layer. The evidence suggests that that may be what happened here, and it's an event that took place relatively recently as the solar system goes, perhaps just a few hundred million years ago. That means for most of Earth's history, the sixth planet from the Sun looked much more ordinary than it does today.

As you get closer to the rings, you have a decision to make: Over, under, or through? You're feeling adventurous today, so you decide to go through the midplane of the rings to start. You accelerate a little to reduce the relative motion between yourself and the nearest particles (isn't orbital mechanics fun!) to less than a meter per second.

It's at this point that you come upon the first particles. They're nothing to write home about—just icy grains of sand and dust. But gradually their number and concentration pick up as if you're slowly being swallowed by a snowstorm. You start to see the odd larger snowball or boulder mixed in among the small stuff. Those are best to dodge and, luckily for you, on very small scales the physics of avoiding objects is much more intuitive. Thrust away from things you want to avoid (point the engine toward them), and toward things you want to investigate (with the engine behind you).

The feeling of suspension is strange and you're constantly having to remind yourself that you are not flying through an earthly snowstorm. As you pass a larger, slowly spinning, ice boulder you reach out and touch it, shoving it gently as if to prove that it is real and not just a figment of your imagination. The sides of the particles are pitted and faceted with ice dust stuck in hollows, the scars of hundreds of millions of years of slow-speed collisions with one another and exposure to space.

But you quickly tire of pushing through this crowded highway and decide to get out of the rings for a bit. It doesn't take much; just a few meters below the plane you exit into a half-lit world. Saturn is still there, just as big and just as bright, shining in reflected light from the Sun. But you know you are on the shadow side of the rings from the diagonal shadow that the rings cut across the planet's yellow disk.

On this side of the rings, away from the Sun, most of the ring particles are themselves in shadow too—silhouettes against Saturn's bulk. Some light does get through, but it's much like lying beneath a leafy tree. Here sparkles from the Sun make their way to your eyes from time to time, but more often you are drawn to the patterns of light and darkness that dapple you and your surfboard. The patterns reel and change as the ring particles move in their slow dance, sometimes blocking the Sun, other times revealing it. Meanwhile, there are enough fluffier snow-like particles to give an overall light moonshine-like cast to the experience.

Curious now, you pass through the rings again toward the lit side. From this vantage point, the rings look like a heaving ocean full of pack ice. But pack ice makes a lot of noise, and this churn is soundless. The Sun is mercilessly bright up above, casting harsh shadows across the bergy bits all around you.

As you gaze into the distance and use a device in your helmet to magnify the view, the oceanic impression is reinforced by structures that look like waves. You know that there are density waves that propagate outward like a spiral from Saturn. But a density wave just means that particles get closer together, not that they increase in height above the ring plane—there is no water below the surface you are seeing, just more floating ice particles.

This, then, must be something else. You decide to investigate and, checking your instruments, realize that the waves appear to be growing in height in the direction you are headed. Could these waves have something to do with Daphnis?

As you get close, it becomes obvious. The leading edge of the wave is right where Daphnis is meant to be. To get an aerial view, you move farther off the ring plane until you are tens of kilometers above it. From here you can see that Daphnis is clearly at the leading edge of two waves that look like boat wakes as the shepherd moon travels down the void of the Keeler Gap. One wave, on your side of the Keeler Gap, seems to be falling behind the moon. But, strangely, on the other side the wave seems to be outpacing the moon.

You are puzzled for a few minutes until you recall that the farther you are from Saturn, the longer it takes you to go around in your orbit. So if Daphnis disturbs a particle in a closer orbit to Saturn than the moon, that particle seems to race ahead. If it disturbs a particle that is in a farther orbit, then the disturbed particle seems to fall behind. It's just a little celestial mechanics, nothing more.

Watching the two wakes of Daphnis, you get an idea. The ring particles don't have any propulsion; all they are doing is responding to the gravitational field acting on them. What would that do to you? Eager to find out, you thrust toward the ring plane and slow your speed toward Daphnis, a few meters above the particles. You can see those particles building slowly ahead of you and you reduce your forward momentum and then kill the engine. Just in case nothing happens, you'd prefer to avoid a crash.

But then, as the particles rise, you find that you are rising too! As you travel over the ephemeral mountain in the ring plane your board keeps its distance from the particles, and before you know it, you are over the top and traveling down the other side. Well, what do you know—you actually can surf Saturn's rings!

On the other side of the Keeler Gap, you take a minute at the top of the highest wave and just sit on a boulder. Seems like a nice place for lunch. As you take in the scene, you pause and think about all the behavior you've seen so far today. But there's no time to waste. While you have some slack in your timeline, celestial motions wait for no one. So before long you head off again in search of a propeller.

Halfway to Pan you find one. You had been surveying from high above the ring plane when the classic pattern of two bars swirling slowly around a central hub came into view. You ready your instruments and dive in for a closer look to begin your survey.

What you find is a ring particle that is unusually large—perhaps the size of a moderate office tower—but is not quite a moon. Sometimes these are called moonlets, and what distinguishes them is that they have enough gravitation that they can attract other particles from the ring, creating the classic propeller shape. Already, you can see those particles piling up near the midline of the moonlet, giving it a thick and fluffy waist. Maybe the data that you and your classmates are collecting will help scientists clear up some of the mysteries about these structures.

Once you finish up at the propeller and its unnamed moonlet, you make haste for Pan.

Above the ring plane, you catch sight of the Encke Gap first before a speck in the middle makes its appearance. As you get closer, you can tell that something is not as you expected. Pan seems flattened, with a soft equatorial ridge of small particles, something like a giant ravioli in space. Maybe this is what happens when a bunch of propellers run into one another?

But all speculation is driven from your mind as you give some thought to a nice meal of real food and perhaps a hot shower. It has been a long day spent in your suit, and no matter how much fun it is to live out the cosmic dance, it's always nice to come home.

★★

Out at 9.57 au, Saturn and its moons are the gift that keeps on giving.[1] There are many interesting things to talk about, so much so that we have dedicated no fewer than four chapters of this book to the remarkable things we find in the Saturnian system, beginning with what is arguably the solar system's most iconic feature: the rings of Saturn.

### WHAT ARE RINGS, ANYWAY?

Who would have thought planets could have rings? They certainly surprised Galileo when he first saw them through a telescope, and in the end their true nature eluded him.[2]

Even with as much as we know about the rings today, they still are odd, surreal, and even alien, a perfect symbol of the unknown and of adventure. But what are planetary rings anyway? Is this normal? Should we expect rings everywhere? Or is Saturn unique? These are all great questions!

As our Saturnian pilot explains, the rings of Saturn are composed of billions upon trillions of tiny individual pieces of water ice, each one on their own individual orbit around Saturn. We can tell what the rings are made of by measuring the light that bounces off them. From 2004 to 2017, the Cassini spacecraft orbited Saturn,[3] measuring a variety of things, including sunlight bouncing off the rings. It then took that sunlight and broke it up into a rainbow, known to scientists as a spectrum. The materials within the rings will interact with the light when it reflects, leaving its mark as small absorption features, a characteristic fingerprint of the chemicals within the rings.[4]

Detailed analysis of the spectra of the rings of Saturn have shown that upward of 95 percent of the rings are pure water ice, with little contamination of other ices (such as $CO$, $CO_2$, $NH_4$, or $NH_3$), salts, rocks, or metal substances. This is an impressive feat, as other water ices in the solar system (such Europa's surface, discussed in chapter 3) are contaminated with all sorts of impurities. It is not too much to say that Saturn's rings represent the purest water in the solar system.[5]

WHAT IS THE DISTRIBUTION OF SIZE?

You would think that after hundreds of years of observing Saturn with telescopes on Earth and sending four very successful missions to and by the ringed planet (Pioneer 11, Voyager 1, Voyager 2, and Cassini), we would have been able to directly image the individual pieces of ice in the rings, and thereby get a direct measurement of the size of the ice chunks. Unfortunately, resolving the ring pieces is not as easy as it sounds. Terrestrial telescopes just don't have the resolving power to discern objects smaller than a few hundred kilometers or so at that distance. Happily, that creates an upper bound for us: we can confidently say the particles of ice in the rings of Saturn are much smaller than a few hundred kilometers! But, on the other hand, this means we must get close to the rings to resolve them.

We've made some pretty close approaches over the years. Voyager 2 flew by Saturn in 1981, at a closest approach of about 100,000 km,[6] but still not close enough to resolve the ring pieces. The Cassini spacecraft spent most of its thirteen years orbiting Saturn much too far away to resolve the ring pieces. However, as the mission neared its ending, NASA directed Cassini to fly through the 10,000-km gap between the cloud tops of Saturn and its rings. It did this twenty-two times before the mission ended, and the closest the spacecraft came to the rings during those orbits was 3,810 km.[7] Yet, even at this distance, we were still not able to resolve the ring pieces individually! Are they the size of sand grains or mountains? The size of cars or cities? The size of pebbles or baseballs? How do we answer this question?

To determine the size of the ring pieces, we use an indirect method called radio occultation. In this approach, a spacecraft, like Cassini, is directed into a position where the rings are in between itself and Earth. It then sends radio light signals through the rings toward Earth. As the signals go through the rings, particles of various sizes will deflect or scatter radio light of different wavelengths. This is demonstrated in figure 6.1. At Earth, we measure what radio light we receive and compare it with the radio light

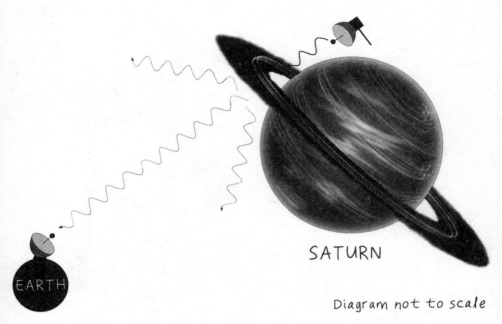

SATURN

Diagram not to scale

Figure 6.1
To perform an occultation, the spacecraft selects a part of its orbit where the rings of
Saturn come between its transmitter and a radio telescope on the Earth. Different sizes
of particles interact in different ways with the spacecraft's radio transmission. Some
allow the beam right through; others scatter the beam in all directions. Based on how
much of that radio signal makes it through the rings at different frequencies, scientists
can reconstruct the sizes and abundances of the particles within the rings.

we know left Cassini. From this, we can calculate what size, or distribution
of sizes, of particles is needed to create the scattering of light we measure.[8]

It may not surprise you to learn that the ring particles are not all the same
size. In fact, simulations suggest that the best fit to the radio occultation
data is a distribution of ring particle sizes, ranging from the very small (less
than a centimeter) to roughly the size of a house at the largest. However,
they are not distributed equally. The data suggests the smallest particles
outnumber the largest particles by about 10 billion to 1! The distribution
in sizes follows the relationship shown in figure 6.2.

Looking at this diagram, it's clear that if you were to fly through the
rings, the particle size you would see most often is something like a snow-
flake or smaller. As the particle size gets bigger, you see fewer of them, to
the point where particles the size of your fist you see frequently enough,

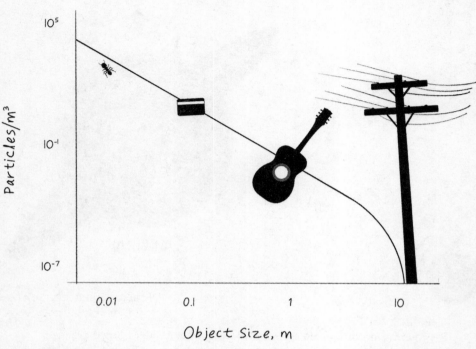

Figure 6.2
Saturn's rings contain many more small particles than large particles. This diagram
shows how many particles of each size are present. Data source: N. Brilliantov,
P. L. Krapivsky, A. Bodrova, and J. Schmidt, "Size Distribution of Particles in
Saturn's Rings from Aggregation and Fragmentation" (2015), https://www.pnas
.org/doi/full/10.1073/pnas.1503957112.

particles the size of a car would be rarer and easy to avoid, and particles the
size of a house are few and far between.

### WHAT WOULD THE RINGS LOOK LIKE FROM WITHIN?

Now that we know how big the ring particles are, we can try to picture
what it would be like to fly through them. This requires knowledge not just
of the size of the ring particles but of how densely packed together they
are. To know this, we rely on a very important measurement that can be
found in many areas of science: *optical depth*.

Optical depth is a measurement of how much light can make it through
a material without being deflected or absorbed in some way. It is measured

as a logarithmic value, where an optical depth of zero means no light is being blocked, an optical depth of 1.0 means about 30 percent of the light is being blocked, and an optical depth of about 2.5 means most of the light is being blocked.

The optical depth of the rings of Saturn can be measured, again, using occultation, though it doesn't have to be limited to radio occultation. Cassini could measure the amount of light from Saturn itself blocked by the rings, or background stars. If we measure how bright a background star appears to be before and during being occulted by the rings, we can measure exactly how much light was blocked.

As may be apparent by just looking at the rings, the optical depth changes across Saturn's complex ring structure. In the A ring, where our story takes place, the optical depth is about 1.0.[9] That means that about 30 percent of the light is being blocked by the ring system. Thus, if you were inside the A ring, you could still see out above and below you, and still see the perhaps ghostly image of Saturn hanging above you. This would be like sitting under a few shady trees and looking up at the blue sky beyond.

But is there any place in the ring structure you could go where, once inside the rings, you wouldn't be able to see out? The answer is yes! In the B ring, the optical depth is about 2.5, which is enough to extinguish pretty much all direct light from the outside world. If you flew your spaceship there, in the middle of the rings, it would feel like you were in a milky half-light illuminated only by light that had bounced off ring particles, or perhaps like you were scuba diving in very murky or deep water after kicking up the sediment.

## WHERE DID THESE RINGS COME FROM?

Are rings normal? Should we expect rings everywhere? Or is Saturn unique? The theories of the origin of ring systems in our solar system are as varied as the multiple locations we find them. All four gas giants have rings, and even some small minor objects have rings. For Saturn, the purity of the water ice found in the ring particles is one of the most telling clues as to their origins.

The rings of Saturn likely formed when a large icy moon got too close to the planet and was torn apart. As discussed in chapter 3, icy moons are rather common in the outer solar system. Europa is a moon with a surface made entirely of water ice, below which there is a liquid water layer on top of a predominantly rocky mantle and core. If a former moon of Saturn, say, an object like Europa, found itself getting too close to the ringed planet, the gravity of Saturn would start to tear it apart, starting with its water ice and liquid water layers. Think of this like tidal forces in the extreme (remember chapter 1?). If the water ice was torn away first, and the rocky core then collided with Saturn, it would leave you with a bunch of icy objects in orbit around Saturn, without any contamination from rocks.

The truth of the matter is the origin and evolution of Saturn's rings are still being debated in the literature. While the composition and size distribution within the rings is well supported by Cassini data, scientists have yet to agree on how the rings formed, and, what's more hotly debated, *when* the rings formed, and how they evolve through time.

As mentioned above, the Saturnian system is rife with interesting places. So we decided not to go very far for our next daydream. Titan, the largest natural satellite of Saturn, is home to some truly interesting curiosities.

With a push on the dock, your sailboat glides gracefully into the wine-dark liquid and you are underway.

Of course, here on Titan, that's not water in which you are floating, but a combination of liquid ethane, methane, and nitrogen. The precise mixture of the three can vary from place to place and with the seasons, which keeps boating interesting on Titan, despite the generally constant calm winds and near lack of waves of any serious height.

Speaking of the wind, once you are clear of the dock it's time to hoist your sails and set your course. What the wind lacks in speed, it makes up for in density, packing a powerful punch with more than four times as much mass per cubic meter as sea-level air on Earth. As your jib and mainsail begin to gather that wind effectively, you can feel the acceleration. It's not long before you build up a good clip. In this direction you are on a run with the wind behind you so have deployed both sails on opposite sides of the boat to gather as much wind as possible in what even an earthly mariner would recognize as a "wing-on-wing" configuration.

Just like back home, the wind filling those sails is mostly nitrogen gas, with about one molecule in twenty consisting of methane and a pressure 50 percent higher than what you're used to. That means that, though you are sailing on an alien ocean nearly 1.5 billion km from home, and the temperature outside is −180°C today, you don't need a spacesuit. Instead, the highly insulated parka with built-in heating and an oxygen mask are all the gear you need. That makes it easy to travel the deck: bending around a shroud here, making adjustments to the rigging there, each move completed through well-gloved hands.

Of course, the light gravity doesn't hurt your mobility either, when it comes to moving about the boat. The enhanced density of the atmosphere combined with the low gravity makes it so you think you could fly away. Indeed, you probably could, which is why you've been careful to tether yourself to the mast with a trapeze harness. You're skilled enough to know that you can jump fluidly to the top of the mainmast in a couple of bounds to undo a tangle if you must. But you don't dare allow yourself more experimental flights. It wouldn't do to fall into the hydrocarbon sea below.

If, in fact, you were to go over the side, all your advantages would be lost. The density of the sea is quite low, at just 650 grams for each liter you scoop up. Since you are made up mostly of water, which has a density of 1,000 grams per liter, that makes it awfully hard to float. That applies not just to sailors but to their boats as well. Even now, your trained eye has yet to get used to just how low your sailboat rides in the liquid and how broad and deep the boat needs to be to generate the necessary buoyancy to keep afloat.

But your sailboat has an extra trick up its sleeve. The keel has several attached hydrofoils, and as you pick up speed the boat begins to rise out of the hydrocarbon soup. In the low gravity, it doesn't take much speed to generate enough upward force on the hydrofoils to get clear of the liquid and into a regime where buoyancy just doesn't matter.

As you skip along the surface you steer your boat toward one of the most interesting features here on the Ligeia Mare: the so-called magic islands that seem to appear and disappear. Are these features made from bubbles of nitrogen? Or are they concentrations of small waves excited by variations in the wind, like an oily version of a cat's paw ripple pattern? They come and go so frequently that no one really knows.

That's half the fun, here on Titan. Because the atmosphere is so opaque in the colors of light we can see with our own eyes, the "images" from orbit come from spacecraft able to see in parts of the spectrum that we cannot. To those spacecraft, the atmosphere looks clearer, but our brains play tricks on us when we try to understand the data they capture. For instance, some of the most crisp and detailed information comes from radar data. But what appears dark and light to a radar instrument is just rough and smooth to human eyes. Complicating matters, radar instruments need to supply their own source of radar waves, like examining a planet under a headlamp.

The seas on Titan were discovered because of their incredible liquid flatness—their surfaces reflected all the radar waves away from the spacecraft into empty space and made the seas look an impossible inky black. Solid rocks and soils instead tend to have lots of imperfections, which scatter at least some of the radar beam energy back toward the spacecraft, and so they appear in shades of white or gray.

So how can an island come and go? Well, you can trick radar if you dapple an otherwise flat surface with something to scatter that radar light back to the spacecraft. That scattering of the radar light will be especially strong if that something is close to the size of the radar waves, in the case of Titan, a few centimeters. It turns out that's just the right size for a froth of bubbles or some light ripples. When the bubbles or the ripples appear, they look just like rocks and soil to the radar, showing up in shades of gray. But when those bubbles or ripples subside, the surface goes back to a mirror-like flatness that looks flat black to the radar. Magic!

As you travel along, you gaze out toward the horizon. That horizon is close here on Titan, nearly twice as close as on the Earth. As a result, even though none of the lakes or mare on Titan are very large compared with similar liquid bodies at home, it is much easier to get lost. From the crow's nest of your sailboat, some ten meters up, you can see only some seven kilometers in any direction—and that's on a clear day. Today you've got a steady snow made up of benzene particles and the complex fractal shapes of the refractory hydrocarbon dust called tholin that gives the Titanian atmosphere its yellow color. This gunk is already starting to form a slick surface layer that your boat's keel must cut through when you are low in the liquid.

You're thankful for your instruments to fix your position. Navigating in any other way would be impossible. With too much haze between you and space, you could never tell your direction by the stars, as would an ancient mariner, at least not with your own eyes. In fact, you can barely tell your cardinal direction from the slightly brighter part of the sky back-illuminated by the Sun. A sextant would be useless on Titan. Overall, the dimness of the scene out here, lit up by less than 1 percent of the light on Earth, means that you tend to favor night-vision goggles for any technical tasks.

But you don't need any navigational aids to identify the "magic islands." Up ahead you see a bright patch of sea, slightly flatter than the liquid around it, with foam on top. To slow down, you release the main sheet, depowering your mainsail. As the boat slows and drops back down into the sea from the hydrofoils, you gain better control over the rudder and your craft's direction. Finally, you heave to, dropping the mainsail just a touch to bring yourself to a standstill.

You take a second just to sit in the silence, feeling nothing but the gentle bobbing of the hull. You reflect that so many epiphanies of nature feel passive, simply being in the right place at the right time to observe some beautiful celestial alignment or phenomenon. But sailing isn't like that. Instead, it is an experience where you feel as if you are working actively and cooperatively with nature. The environment provides the wind, and with nothing more than a bit of skill and some well-designed bits of fabric and fiberglass you can generate the pure joy of motion, tapping into some deeper, inexplicable connection.

A bubble bursts nearby over the side of the boat, breaking your reverie. Peering over the rail, you can see the surface manifestation of your quarry. It's almost as if there were titanic whales down below bubble netting for fish. Perhaps today you will learn the secret of the "magic islands." From the cabin, you pull out a hydrobot about the size of a dog. This swimming robot has all the instruments you'll need to investigate this phenomenon and can even bring back samples for analysis. You drop it into the water with a "bloop" and let the onboard AI take over, executing the program you wrote onshore. The hydrobot dives away.

It's long been known that Titan has all the molecules needed for life. Indeed, this is a wonderful place to study what might be the precursors to the active biology we see in other places, like the Earth. Those molecules are constantly combining, breaking apart, and recombining with one another, like dancers trying out all the possible arrangements of partners and movement. But because of the intense cold, they do this very slowly. As a result of this slow-motion chemistry, the moon is locked in a pre-life, or "pre-biotic," state.

Was it always like this, and is it still pre-biotic everywhere, even now? Certainly, as you go deeper Titan must get warmer. There are even indications of liquid water beneath the surface—actual water made of two hydrogens and one oxygen—as the main material making up the moon melts under intense heat and pressure. Could there be someplace, in some crack or cryovolcano, linking the world of warm liquid water with the world of cold chemistry, where, as they say in the movies, "life finds a way"?

Ultimately, you prefer not to speculate. Data is what you are looking for and, indeed, you can see Data coming toward you now as the hydrobot breaks the surface on its return leg. By "Data" you don't just mean the samples and observations onboard the hydrobot; it's also the name that you gave to the hydrobot itself. As a scientist, you just can't help an obscure reference to an ancient piece of science fiction.[1] You chuckle inwardly as you reflect that you and your kin in this profession are certainly strange folks.

You pick up Data the hydrobot out of the liquid hydrocarbon and clean the robot with a towel. As you place it in its alcove below you take a minute to examine the samples from the murky bottom, holding them up to one of the boat's lights as if you could divine anything in that way. Maybe in these tubes there lies something more than just long-chain hydrocarbons. But that analysis will have to wait for later.

The recording of the robot's journey is fascinating, however. You watch through its eyes at triple speed the descent through various sharp-looking interfaces as the composition of the liquid changes. It's as if there were multiple haloclines, wavy curtains through which the robot needed to pass before arriving at the not quite liquid, not quite sediment of the bottom that is happily bubbling away. One of your students will be thrilled to pore over every detail.

For now, however, it's time to move on. You grab the halyard and hoist the mainsail. Once you're out of the water, you gradually turn 90 degrees and trim your sails for a beam reach. You're heading toward a peninsula you know is just over the horizon. You can be forgiven for taking a bit of time to explore the coast. It's not as if the detour will cost you any fuel.

Additionally, you're a firm believer that it is one thing to explore a place from a satellite or on a map and something very different to do so with your own eyes. There's even something different about the perspective from the sea versus hiking on a headland. Besides, you tell yourself, you might see some interesting phenomenon whose scale is so small that it's hidden from those eyes in the sky, perhaps something for your next research proposal.

But really all this tooling around is just for fun. You spend a couple of hours in this way, tacking back and forth exploring several bays and inlets. You don't even care if you find anything at all.

Still, you can't do this forever, and, gradually, you notice that the character of the light starts to change. You call up a weather map and notice a strong convection cell is headed your way. It's time to make for your boathouse back on the lee shore with all speed.

Squalls on Titan are nasty affairs. Sometimes the convective cells can be stratified into several layers, with thunderheads on top of thunderheads like some child's fever dream of an impossible storm made real. Those storms can drop as much as a meter of methane rain all told, a mind-boggling amount for someone used to terrestrial storms.

You don't want to get caught out in that! With some urgency, you adjust your rigging to start beating into the wind. From here it will be a race with the storm to see who gets to the coast first. In some slightly deranged way, you're looking forward to the passage.

This type of sailing is always exciting, and it's rare that you get to do this in such a well-behaved atmosphere as exists on Titan. On the way out, you were moving in the same direction as the wind, so everything on the boat appeared calm, like in a hot-air balloon. Now, close-hauled and sailing into the wind, the speeds add up until there is a veritable howling.

In the open sea, you can see the storm now looming over the horizon. You can even see flashes of lightning, an unusual sight, like something alive and malevolent. As if aware of your existence, the winds increase and gust as you begin to catch the outflow from the storm. But the extra wind just powers up your sails. Quickly, you reef them, reducing their surface area to compensate. It's all you can do to keep the boat from flying completely free of the sea and dashing itself to bits on the splashdown or on an errant rock.

As you pull up to the dock, the first droplets start to hit the water and the wind calms just a little as if the storm knows it has been beaten. Just in time, you think, as you drop the sails completely and fold down the mast. It's all you can do to keep from vibrating with the exhilaration! But as much fun as that was, there was a taste of danger in there as well. Today was a reminder that nature's power is to be disrespected at our peril. Though we are space travelers, none of our technology can invalidate that fact.

### HIDDEN LAYERS IN TITAN'S LAKES

When we began our exploration of the rings of Saturn in the previous chapter, we were floating just outside the A ring at 76,000 km from the cloud tops. In this chapter, we've moved to Saturn's largest moon, Titan, at 1,221,870 km from the planet.[2] Titan is one of the most interesting and unique objects in our solar system. In fact, we felt it was worthy of two chapters in this book (make sure you grab a coffee for chapter 14). Its characteristics almost make it feel as if it should be a planet on its own, rather than classified as a natural satellite, or moon. Of course, being in orbit around Saturn means it is, by definition, not a planet, but many comparisons can be made to other unique places, like Earth, for example.

First, Titan is the only moon in the entire solar system that has a substantial atmosphere. There are some places, like Europa, that have exospheres (see discussion of exospheres in chapter 4), but an exosphere, while made of gas, is not the same as a fully-fledged atmosphere. The main difference between the two is that, in an atmosphere, the gaseous molecules interact with each other, whereas in an exosphere, the gaseous particles are so spread out that they don't often encounter one another. But not only does Titan have an atmosphere; its atmosphere is denser than our own, by about 1.5 times. A denser atmosphere than Earth . . . on a moon. Wild!

You might think this could mean the atmosphere is warm (think about the dense atmosphere of Venus we talked about in chapter 5), but Titan is so far from the Sun that the average temperature in the atmosphere is down near −180°C.

Even more interesting, the most abundant gas in Titan's atmosphere is nitrogen ($N_2$), which is also true for Earth, but on Titan, nitrogen makes up about 95 percent of the gas, compared with about 79 percent on Earth. After nitrogen, the next most abundant gas is methane ($CH_4$). But due to the high pressure and very cold temperatures, methane can condense into a liquid, leading to methane rain and methane lakes.

Yes, you read that right: there are lakes on the surface of Titan! Pretty big ones too, rivaling Earth's Caspian Sea at the largest. Don't get ahead of yourself, though; these are lakes of methane, not water. Still, there are only two places in the entire solar system where there are sustained pools of liquids on the surface, and they are Titan and Earth. See what we mean? Titan is unique.

How do we know these lakes exist? It's not as easy as just seeing them in a picture taken from an earthly satellite or one of our robotic emissaries that has visited the Saturnian system. This is because Titan's thick atmosphere contains a relatively large amount of methane prone to creating a thick, yellow-orange haze made of a group of chemicals called tholins. In general, tholins are a variety of organic compounds made up of nitrogen, hydrogen, carbon, and oxygen; that appear yellow-orange; and that can float around as aerosols in Titan's atmosphere. The haze they create is so thick that if you were in a spaceship orbiting around Titan and you looked down at the moon, you would not be able to see the Titanian surface.[3] So what do we do? The same thing we do everywhere there's a gas we want to look through but can't: we send radio light through and look for a reflection.

The Cassini mission made frequent passes of Titan during its 13-year mission at Saturn, each time sending radio light through the haze to measure the surface below. This mapping of the Titanian surface has revealed some interesting topography, including mountains, channels, plains, and all manner of other geological formations. But if you remember our discussion of radar mapping of Venus in chapter 5, radio light bounces off different materials in different ways. At Titan the reflections of radio light uncovered large sections that reflected the radio light away very differently than if it were bouncing off rock, leading us to discover giant lakes of methane.[4]

But we don't think it's just liquid methane in these lakes; there are other chemicals that, with the low temperatures and high pressures of the atmosphere, prefer to be liquids rather than gases. Ethane is one example. Where methane is the simplest hydrocarbon, composed of one atom of carbon combined with four hydrogen atoms, ethane is the next most complicated hydrocarbon molecule ($C_2H_6$). Because it is a little bit bigger and a little bit heavier, it prefers to be in liquid form even more than methane. We have even seen cases where ethane seems to be the main component of a lake on Titan.

But for most of the lakes on Titan, there is likely a complicated soup of different hydrocarbon liquids along with some solids and even dissolved gases. When you mix multiple chemical species into a single liquid, strange things can happen. Just think of the polar oceans on Earth that can remain liquid even in bitter cold due to the amount of dissolved salt. In the same way, different conditions of temperature, pressure, and composition can change the properties of the liquid making up Titan's lakes and seas. In some cases, this can result in the mixtures separating out into two different liquids that lie one on top of another.

If you were swimming in one of Titan's seas, you would likely be able to see a change in optical properties between the layers. The same thing happens on Earth when fresh and salt waters meet. If the conditions are right, the different densities of the two water bodies can keep them separated by a sharp boundary called a halocline. Because the two layers transmit light in different ways, the halocline sometimes even looks like a reflective surface in diving pictures.[5]

Over the course of the Cassini mission, a variety of very large seas (called mare) and smaller lakes (called lacus) were catalogued, including three very large seas: Kraken Mare, Ligeia Mare, and Punga Mare.[6] In Ligeia Mare, the second biggest sea on Titan (about the size of Lake Superior in North America), the data suggests something very odd: what appeared to be a solid portion of the shoreline that created a peninsula had disappeared suddenly, only to reappear again in later radar images.[7] This phenomenon has been seen a few times, including in the largest sea, Kraken Mare, where islands magically appeared and disappeared through multiple radar soundings.[8]

Of course, magic is just science we don't understand yet. Is it possible this is related to flooding, where large portions of land are being continually covered and uncovered by liquid? Maybe, but probably not. These "magic islands" are probably best explained by the mixture of gases into the various liquids, and gases like nitrogen coming out of solution as the conditions at depth change. Think of it like opening a bottle of soda pop— the sudden change in pressure releases the dissolved carbon dioxide in the liquid. And, like the foam that forms on top of that soda, the nitrogen bubbles would give the liquid surface some texture. That, in turn, would make these areas look more like land to radar than like a flat and reflective liquid surface.

## SATELLITE STORMS

Titan's surface is largely shielded from solar energy by the dense hazes present in the satellite's mid- to upper atmosphere, beginning at 60 km above the surface. That means that typical winds on Titan near the surface are light and often blow in the same direction all the time. This creates very little temperature contrast from pole to equator. It may often only be a degree or two Celsius warmer in the tropics than at the pole. That would give someone less incentive to fly south for the winter on Titan.

But this doesn't mean that storms don't happen on Titan. Sometimes you can even get convective storms, the kind that we call thunderstorms on Earth. We can often see evidence of these storms as bright clouds whose tops climb into the upper atmosphere. At other times, we can infer the existence of these storms from the effect of their rain on the landscape or by running complex numerical simulations with computers on Earth. The latter are a bit like weather forecasting models, but we run these to try to

Figure 7.1
Titan's complicated soup of hydrocarbons and dissolved gases can take on many different forms depending on the conditions of pressure and temperature. In some cases, those gases can come out of solution to form a frothy foam on the surface that can make spacecraft think an island is present.

match the conditions we see instead of trying to predict what will happen in the future.

Those inferences tell us that storms on Titan can be truly spectacular. Clouds can drop a meter of rain during a single storm.[9] Or you can get two different convectively mixed zones on top of one another. That opens the possibility that you might end up with several layers of thunderclouds.

Remember that Titan is the only other planetary body in our solar system with the equivalent of a hydrological cycle. As you might expect, that makes its seas and weather anything but boring!

Earth is not the only planet with similarities to Titan, however. Like Titan, Venus has a dense, optically thick atmosphere that rotates faster than the surface turns. We'll take a closer look at this high atmosphere in the next chapter.

AROUND THE WORLD IN EIGHTY HOURS

It's still dark outside when you wake up. The generously sized porthole in the wall is simply a deeper shade of pitch, revealed only by a patchy sprinkling of stars. The starfield as viewed from your hotel room is not inherently patchy, but the missing pinpricks of light are elided by the unseen black of Venusian clouds. Thicker than the clouds of the Earth and without even so much as a moon to provide light for their sulfuric acid particles to scatter, they hang, ominous by their absence in the early morning sky.

In little more than a few hours, you will be getting up close and personal with those same clouds. You hope that they are friendlier by day than they seem right now.

Turning your gaze from the porthole, you glance around the room. You paid for the view, as many tourists do. The room is spartan but comfortable; if you close your eyes, you could even believe you are back on the Earth. The gravity is the same, and so is the pressure—certainly no worse than a night spent at altitude. While you wouldn't care to open the porthole to let in the carbon-dioxide air of Venus and the acidic humidity, the window is clearly meant mostly to be weatherproof and air-tight rather than holding any significant pressure in or out.

The reason for all of this is that up here in the clouds, 50 km or so above the hellish surface of Venus, your hotel is gently floating in what might be the most Earth-like environment in the solar system. Gravity, pressure, and temperature are all close to the values you know and love. Plus, the air within the building is naturally less dense than the surrounding atmosphere of Venus. As a result, every single cubic meter of open space within the structure provides about 600 grams of lifting power.

This encourages wide-open spaces in the architecture of floating Venusian cloud habitats. Couple that with the need for thin, lightweight, and acid-resistant membranes, and the result is a bulbous plastic structure that, from a distance, doesn't look all that different from a cloud itself! Truly an inflatable castle in the sky. Or a sky whale, depending on your choice of metaphor.

You arrived a few days early, just to explore this kind of architecture. While early versions included separate balloons tethered to the main structure, modern cloud habitats include them directly in the design. This allows one to walk out beneath the stars or to take in the drama of a spectacular Venusian sunset over a meal with friends while appearing to be outside. There's even a clear-bottomed pool for the adventurous who can imagine themselves flying through the clouds with each dive. But such a mass of water in a floating structure seems like a terrible luxury—nearly a contradiction in terms—to an aircraft designer such as yourself.

All this takes place beneath a gentle, transparent canopy made up of individual gores whose sutures are lost in the distance to the casual resort guest. Thankfully, the membrane can also go somewhat reflective through some electrical process to keep out the worst of the punishing Sun, which delivers twice as much heat here as on the Earth at high noon.

Naturally, the structure has some propulsion to deal with weather, and ballasting to keep it steady and, when necessary, to control its altitude. There may even be some gentle swaying, you think as you lie in bed. But that might just be your imagination deceiving you as you drift back to sleep.

A few short hours later and you have built up too much excitement brought on by nervous anticipation to sleep any more. You are now busy going through your checklist, sitting on your bed and taking stock of each and every piece of kit. Most of the items are small tools, and there is a backpack of foodstuffs that you have carefully prepared.

But you are most proud of your outfit. Most of your fellow ballooners at this festival have opted for modern fabrics and seamless design, which makes them look like elite athletes or astronauts.

Not you.

You've always been a fan of anachronisms, favoring glasses to treat your myopia rather than the quick and easy surgeries preferred by others. Wearing a mechanical watch instead of a modern digital chronometer. There's simply something romantic, you feel, about the trappings of the early era of aviation and ballooning. You have already put on the long pants, a light green peacoat and tall leather boots. As you stand, you pick up a plain leather cap and goggles, and grab a long, insulated, brown leather jacket lined at the collar with fur. Of course, these garments are not made from actual leather or fur, but are clever imitations that are much better at regulating your temperature and moisture.

Looking in the mirror, you feel you are the very model of the classic aviator, and everything, having been designed just for you, fits you perfectly. You allow yourself a smile. Frankly, if one is going to do something as patently ridiculous as ballooning on Venus, you might as well look the part.

Then you add two extra touches. The first is a breathing mask to provide you with fresh oxygen from a cylinder that fits within the outer coat (you are ballooning on Venus, after all). Finally, you add a striking and colorful scarf made of some silk-like material. You feel that in the scarf you may have improved upon the historical model.

Satisfied, you grab your pack and head up to the launching deck.

This hotel caters to balloonists and those who want to watch them launch or take short rides. As a result, the deck in the center of the structure is designed to give as much protection as possible from winds. Furthermore, near dawn—the best time to launch—the hotel is allowed to drift with the air to minimize any shear. On your way up, you smile and wave at a few fellow travelers and look out westward toward the horizon, with the faint pre-dawn glow that tells you the day is approaching.

Across the empty expanse of sky, you see other hotels, much closer than they would typically be. The great floating structures have gathered here for the same reason that you have—the annual Venusian Balloon Festival, the largest event of its kind in the solar system. In just an hour or so, nearly 1,000 balloons will take flight together and begin an around-the-world tour.

Such a tour is possible because of Venus's super-rotating upper atmosphere. While the rocky part of the planet rotates very slowly, completing a revolution once every 243 days, the upper atmosphere travels much more quickly, completing a circuit in just a few days. This happens because the clouds of Venus are so thick that most of the light from the Sun never makes its way down to the surface. All that energy deposited at high altitudes must go somewhere, and that means spinning up the clouds and upper atmosphere to terrible speeds of hundreds of kilometers per hour.

These winds are the fastest in the southerly mid-latitudes and fall off to the north and the south. Even though all that can be seen of Venus from a telescope or from orbit is clouds, at least in the colors the human eye can sense, you could pick out this location easily. If you've ever seen a picture of Venus, you've probably seen something that looks like a sideways chevron or an angle bracket. The pointed end of that feature is where you are now, with the slower winds falling away behind on the globe at latitudes below and above.

Though the winds here are as fast as any terrestrial hurricane, the cities in the sky move with the winds and are swept sedately around the planet every few days. You wouldn't even know you were moving, were it not for the Venusian satellite positioning system. But the motion of the atmosphere is a good thing, giving rise to the sunset and sunrise cycle observed from your hotel and allowing journeys to any point on the surface near the latitude line of the hotel. Entertainingly, because both the surface and cloud-tops rotate in the opposite direction to those of the Earth, the Sun appears to rise in the west and set in the east. The brochure in your room made light of this fact, insisting that a vacation on Venus was so restful it could turn back time.

You arrive at the launch deck. Placing the goggles over your eyes and checking the seals and flow of oxygen through your regulator, you walk through the lock and out into the open air. Everyone around you is making their own preparations, the entire deck a hive of activity. Because there are so many more balloons than normal here, the very skin of the hotel, which curves gently upward away from the deck in all directions, has been conscripted to help. Stairways and smaller temporary decks have been placed upon these transparent hills to accommodate individual fliers.

As your balloon is for a single person only and is on the smaller side, your deck is located near the apex. You climb the stairs to your berth and begin the process of getting the balloon ready. First, you confirm that all the materials you need are here and then check the lines. Your balloon is of a very simple design. A standard Montgolfier-style envelope encases air that can be heated by a hydrogen/oxygen burner. You can therefore control your altitude like any regular hot air balloon—more heat makes the balloon rise, while laying off the gas causes the balloon to descend.

To save on fuel by providing a baseline of buoyancy, your balloon envelope has a built-in toroidal cell called the "donut" that can be filled with helium. Because of the density contrast between carbon dioxide and helium, that torus doesn't need to be huge, and yours is sized for Venus. Even though it isn't quite historically accurate, you're happy to have this help—without it, an equipment failure could have you descending through the atmosphere to burn up on the surface below. It also makes it a lot easier to fill the envelope.

Once the donut is filled with helium, the envelope stands upright, though drooping in a way that isn't entirely healthy. However, a short sequence of firing from your burners solves that problem, and it isn't long before the envelope snaps sharply into place. The balloon might even ascend of its own accord if it wasn't tightly tethered to the hotel.

You step into the gondola and your weight helps the mooring cables to hold the envelope down. You take a look around. The gondola looks like it's made of wicker, but the composite fabrics that make up the deck and sides are much tougher and lighter. This mass saving is one of several advantages your balloon has on those ancient flying machines. On Venus, the working fluid, carbon dioxide, is heavier per unit volume at the same pressure than the nitrogen/oxygen blend in the atmosphere on Earth. This means that a smaller balloon is needed to produce the same amount of buoyancy. Also on the positive side, Venus's gravity is a shade less than that of the Earth, so the payload of gondola, passenger, and kit weighs less.

But there are negatives too. Because the carbon dioxide atmosphere is inert, you need to bring not only the fuel for the balloon's burners but the oxidizer as well, in your case, pure oxygen. Plus, you too need to be able to breathe, though that requires a considerably smaller mass than the burners themselves. In the end, the calculation is pretty much a wash and many of your colleagues are using balloons that would fly just as well on Earth as here on Venus.

Your gondola is a bit larger than what is traditionally flown on the Earth. This is for comfort and to accommodate a small, pressurized tent in which you will have your meals and take care of any other pesky human business. The gondola also has slightly higher sides for an extra degree of safety. While balloons move with the wind, such that you rarely would feel a breeze while a passenger, Venus's atmosphere can be unpredictable, especially on the night side. It's not unusual to get surprising updrafts or eddies that can suddenly move a balloon in strange ways. This is not a planet on which you'd like to be pitched over the side.

But you're in little danger of that. Looking around, you can see that most of the other balloonists have completed their pre-flight checks. Indeed, the whole flotilla has arrived at the exciting pre-flight moment, just waiting to be released into the sky. It wouldn't make sense for all the balloons to rise at once, and indeed, there is an order—a method to the madness. You signal the air traffic controllers that you are ready to go and are rewarded with a departure sequencing.

And then it begins.

One by one, the balloons alight into the air. The smallest and lightest first, rising quickly like champagne corks seeking higher altitude. The lumbering beasts and dirigibles will be the last to leave the nest. Since your craft is on the lighter side, you guess that your takeoff moment will come early. You are not disappointed: the departure indicator lights up green shortly after the first balloons head skyward. In a well-practiced flash, you move to disengage the balloon's tether and you are away.

While your balloon is not the quickest to rise, the hotel still seems to dwindle rapidly below. Looking above you, the first volley of aircraft have already found a faster current at their altitude and have started to spread out off to the east, like a giant banner unfurling. Looking off in the distance, you can see the same process happening at the other hotels. It reminds you of dandelion seeds being spread to the wind, or coral spawning in the ocean. As with the group leaving from your hotel, the top-most balloons are sheared away to the east, just like the top of an anvil cloud and for the same reason.

The variety of shapes and forms is spectacular. There are classic Montgolfier-style balloons like yours, as well as many more modern aerostats. Airships and dirigibles of all shapes, colors, and forms are present. One advantage of such craft is that their fully enclosed gondolas allow them to travel to higher altitudes where the air is thinner while keeping their crews in perfect comfort. For those seeking greater thrills, there are part-solar balloons with black-painted tops whose solar heating contributes to their rise during the day and their descent at night. At the extreme end are "kytoons," a hybrid between a kite and a balloon that takes great skill to maneuver. For safety, many of these have tethers to the larger dirigibles.

There are even recreations of the Vega balloons, the first to visit Venus and to explore its atmosphere way back in 1985. Some even date this balloon festival to that original around-the world journey. However, a purist might note that the 60-hour battery charge on the Vegas did not allow them to completely circumnavigate the planet before their onboard computers died. An even greater purist might note that the balloons and their silent payloads likely made many circumnavigations before the gas in their envelopes leaked away and the craft descended to the hellish surface of Venus.

For you, it's the huge gap in years between the Vegas' flight and the return of ballooning to Venus that really adds the equivalent of a special circumstances asterisk to that 1985 date for the origin of this event. Nevertheless, since that resumption, the festival has been held each year. There's just something that draws the mind, especially those with an appreciation for ancient craft, to flying balloons on another planet, especially one whose entire livable space is made up of sky.

Gradually, over the course of the next hour, the launches end and the balloons spread out. This isn't so much a conscious choice of the pilots as it is a consequence of atmospheric flow. At this southerly mid-latitude, the air gets faster the higher you go. At the 50-km altitudes favored by the hotels, the winds typically travel at about 60 m/s, or just above 200 km/h. That's fast enough to get you around Venus in about 125 hours. By the time you reach the altitude favored by the Vega probes, at 54 km, it's not unusual to see speeds of 90 m/s, or about 320 km/h, fast enough to circumnavigate Venus in just over 80 hours.

That's a typical altitude for you and your colleagues. However, if you could push on even higher, up to 65 km, there is an exceptionally fast zone in the Venusian atmosphere above the cloud deck where winds have been clocked at up to 150 m/s—over 500 km/h. A balloon at that altitude would take under 50 hours to make the journey around the planet. But it would be uncomfortable. The air is chill there, at about −50°C, and the pressure is only a few tens of millibars, conditions that would put you in the stratosphere on Earth.

No thanks, you think. Eighty hours is just fine with you for a transit time. Even up here, the air pressure is more like what you would experience on a mountaintop, and you are thankful for the plentiful oxygen you've brought along for your burners and for your own lungs.

As you arrive at your desired altitude, you check the rigging and switch the burners to an altitude maintenance mode. With everything set, you get an opportunity to look around you. Because you are traveling with the wind, everything seems incredibly calm. Right now, you are traveling through an area of open air, but you know you're unlikely to be able to steer clear of clouds the entire time. So, it's important to take advantage of the opportunity.

Looking down you can see a deck of clouds far below, like some fluffy plain of white.

The hotels are now lost to your vision off behind you. Pulling out a pair of binoculars and knowing where to look reveals them to you just as they slip away over the horizon. You recall how massive these structures seemed when you arrived, but compared with the atmosphere of Venus, they are but motes of dust. And Venus has far from the largest atmosphere even within our own solar system!

Clearly, you and your fellow balloonists are alone now in your collective isolation. But this is not a true wilderness, nor are you ever far from help here. The festival, being a well-regulated annual activity, comes with plenty of safety craft that can help in any emergency situation. Swinging the binoculars around, you can make out a safety officer now, steering an apparently overpowered dirigible from this balloon to that one, checking on everyone. Should the vagaries of the wind blow you too far off course, you know this craft will helpfully tow you back to the group. It's a minor indignity you'd rather avoid. You don't even bother to look up where there are chase aircraft circling at a respectful distance in case of real trouble.

Quickly now, the Sun climbs in the sky overhead. You anticipate about 40 hours of daylight, and you can see the solar balloons rising serenely above the others as if you are gliding along a plane and they are walking a gentle slope. It's that slow pace of things that always attracted you to ballooning, the feeling of traveling while working silently in concert with nature. A pace so smooth it feels like you're not moving at all unless you look down.

Many find the experience of ballooning off-putting. A lifetime spent in the cabins of powered vehicles cutting across the land, through the waves or in the air, has conditioned them to equate silent motion with emergency. Ballooning therefore is a supremely strange experience for many. Although you do not miss the reassuring hum of an engine, you do appreciate the periodic staccato of the burners that tells you that your altitude is right where it should be.

While the air is clear, you take some time for your own pursuits. You have a light meal in your tent. You spend some time studying the clouds. You even indulge yourself in the luxury of a little sketching of the airborne scene. While it's true you could just take a picture, there's something about the act of drawing that appeals to your sensibilities. This year there are new airships in the festival, and you want to get their colors and shapes just right. At home you have a wall dedicated to your favorite sketches from each year.

You can see from the radio traffic that many of your fellow balloonists are using the time to hail one another and to share the excitement of the day. You smile, thinking of your first few festivals. But you are content, here in your own gondola, to just allow this peaceful time, free of responsibility, to flow over you and to enjoy the moment. You even take a few hours to nap—40 hours of daylight makes for a long day.

Unfortunately, experience has taught you that, whether in life or in the skies of Venus, such sunny days can't last.

In the late afternoon, storm clouds loom ahead. You can't steer your balloon around them, so it's simply a question of picking the altitude at which you can face the storm with confidence. You make the best choice you can, borne out of years of hard-won experience, and then clip yourself in. The safety craft do a final check-in to ensure everyone is ready for what lies ahead. Each balloonist will, effectively, face this challenge alone.

Though your balloon and gondola look like antiques, they are made of tough modern materials, and you know that they can handle whatever Venus throws at them. But for you it's going to be a bumpy ride. You add a clear overcoat to your garb and make certain that no skin is exposed as the balloon enters the sulfuric acid cloud.

No matter how many times you experience this transition, it's always disorienting. Where your view previously extended for tens of kilometers downward and hundreds of kilometers along the limb of the planet, now you're in a churning bubble that makes it hard to see more than a few meters. It's much darker here with the cloud particles absorbing and scattering the light. Indeed, you can reckon your depth within the clouds based only on the apparent darkness of your world. That world is punctuated every so often by blinding lightning flashes that seem to come from everywhere around you and that light up your radio with static.

You keep your hand ready on the burners, but at this point there's little you can do.

As that thought passes through your head, you feel a lurch as if the ground has fallen out from under you. You're in a downdraft, and a look at the altimeter tells you that you're losing altitude fast. You would need to drop an awfully long way to have a problem, so you keep your hand steady. Looking to the lines, they go slack, but the distance between the gondola and the envelope remains unchanged. It's disconcerting, and you can sense the other balloons correcting with bursts from their burners. You spare a glance at your radar—a balloon under thrust in zero visibility is more likely to collide with other balloons than if they follow their own air through the cloud. You can't see those craft, but the commentary you hear over the radio tells you some are working hard against their circumstances.

However, you keep your faith in the principle that, in an atmosphere governed by buoyancy, what goes down must eventually come up. After all, even the Vega probes noted especially severe vertical movements during their flight, and they pulled through.

The next thing you know, you are rewarded. The falling slows, stops, and then you feel pressed into the gondola floor as the downdraft gives way to an updraft. You gradually feel the world getting lighter and brighter.

And, suddenly, you are free of the cloud, emerging into clear air with the Sun close to the horizon. The passage through the storm cloud seemed to last hours, but a look at your watch confirms that you spent only a few minutes in there. Our memories sometimes seem to be such unreliable recorders of the most turbulent times in our lives.

One by one, the other balloons pop free of the storm. Some have extra vertical motion that causes them to oscillate around their target altitude like a float bobbing in a pond. At first, they are shiny with sulfuric acid in the low light. But after a few minutes the liquid dries and evaporates. Eventually, even the pilots begin to reclaim their equilibrium.

You would expect the radio chatter to be jubilant with the catharsis of a thousand balloonists exiting from the storm. But instead, everyone is silent, and you can feel everyone staring out toward the sunset. You've all seen this display before from your hotels. But nothing quite prepares you for the sight out here in the open air as the Sun, a slightly larger disk than when viewed from the Earth, plunges toward the horizon.

To be honest, it's not that different from an earthly scene with the blue sky turning to shades of orange, yellow, and red toward the Sun and to deep lavenders, violets, and indigo on the other side of the sky. Without much in the way of dust particles at this altitude, the colors are dominated by the same Rayleigh scattering from molecules you remember from home. The disk of the Sun itself has turned a bright red and the clouds around you have taken on a rosy glow.

When you turn around, there is so much cloudscape behind you that, where the angle is right, you can see rainbows apparently sitting on the clouds. Because you are so close to sunset, you would swear that the red bands are the brightest, but you can still make out all the colors. You know the rainbow continues past red into infrared—orbiters have seen those refractions with their unhuman eyes—and you expect that there is an ultraviolet portion to this rainbow as well. It's just one more reminder that with its spherical liquid droplets, Venus's middle atmosphere is the most similar environment in the solar system to the Earth's.

As the growing gloam begins to extinguish the light, you turn around. While many of your fellow balloonists are transfixed on the disk of the Sun itself, you know that the best part of the show is in the other direction. The wall of cloud you exited is now a canvas on which the progress of the sunset becomes evident. At first, the whole cloud is awash in reddish light. But slowly at first and then faster, a zone of dusky lavender with a clean line separating it from the bright pink above forms and starts to rise up from the bottom of the cloud. The line is essentially a projection of the horizon at that altitude, with higher altitudes of clouds able to see the Sun for a few seconds or minutes longer. Eventually, just the very tops of the clouds are illuminated. Then they too seem to wink out, changing color to a dark lavender.

Unfortunately, at this altitude there is no temperature inversion, so no possibility of the infamous green flash now at the moment of the dying of the light. But that's no great loss as the flotilla prepares for an even rarer sight. Gradually, each of the balloons seems to come alight from within. The Montgolfier-style airships turn on their burners to salute the sunset, giving a fiery glow from afar. To enhance the effect, many balloons have their own internal illumination to light up their envelopes in every color of the rainbow. Some even shift their colors and hues in the solar system's largest "glowdeo."

It's like a neon-electric version of a thousand enormous paper lanterns, floating in the Venusian sky, truly a spectacular sight to behold.

As the dark becomes complete, the light from the airships competes only with the brilliant stars in the moonless sky and the very faint reflections of that starlight from the clouds below and around the flotilla. You can make out one particularly brilliant bluish star rising in the east without a hint of twinkling: the Earth. Sometimes, Venus and Earth are referred to as sister planets. Each is particularly reflective on account of their clouds. It therefore seems especially fitting that Venus is the brightest "star" in the Earth's sky and that the Earth is the brightest "star" in Venus's sky. However, when it comes to observing, Venus has the advantage.

In earthly skies, Venus can only be seen at night within a few hours of sunset and sunrise, as it is closer to the Sun than the Earth and therefore is never far from our parent star. This means that Venus is up in the sky mostly during the day on the Earth—if you know exactly where to look, you can see it with a good pair of binoculars or a telescope. But you must be exceptionally careful not to get the Sun in your instrument or you risk doing serious damage to your eyes. As well, from Earth we only ever see a crescent Venus, as it is illuminated obliquely by the Sun from Earth's perspective.

On Venus, however, the Earth is a nighttime companion and can be more than three times as bright, even though it is a bit less reflective. Tonight, the disk of the Earth is full, rising opposite the setting Sun. If you had the appropriate equipment, you could even make out the continents as well as the oceans, which are as dark as the clouds are bright and reflective. The hotels have telescopes set up just for this, but you've left such things behind on your journey. Still, it is meaningful for you to simply gaze upon that blue speck in the night sky—your home, now so very distant, but still with you.

In some ways, night is a quieter time than the day. Without solar heating driving the engine of the atmosphere, the building of convective clouds stops. There are some scattered storms that can trouble you on the night side, but this year you manage to stay clear of them.

Instead, things get strange. With so much inertia in the moving atmosphere, the strong jets that propelled you this far begin to break down into eddies and waves that can take a balloon hundreds of kilometers off course and drastically cut its airspeed. The lights on each balloon thus serve another function—allowing the safety airships to keep the flotilla well herded during the night. Every so often a balloon catches a stray patch of air and veers away, only to be towed back to the main flotilla.

To simplify things, the safety airships gather the balloons in a tighter formation than on the day side. The Montgolfiers, like your own craft, lacking any control of their flight other than up or down, are right in the middle. You're not close enough to reach out and touch the other balloons. But

Plate 1  Golfing on the Far Side of the Moon

Plate 2  Rappelling into Valles Marineris on Mars

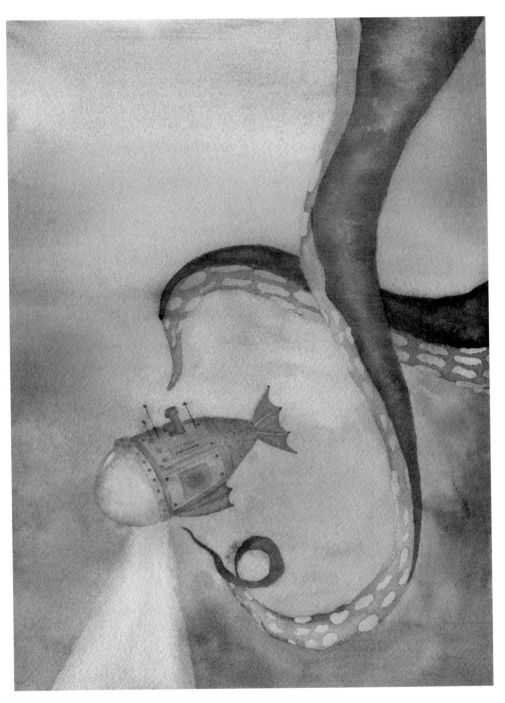

Plate 3   Beneath the Europan Ice

Plate 4   Trekking Mercury's Crest of Dawn

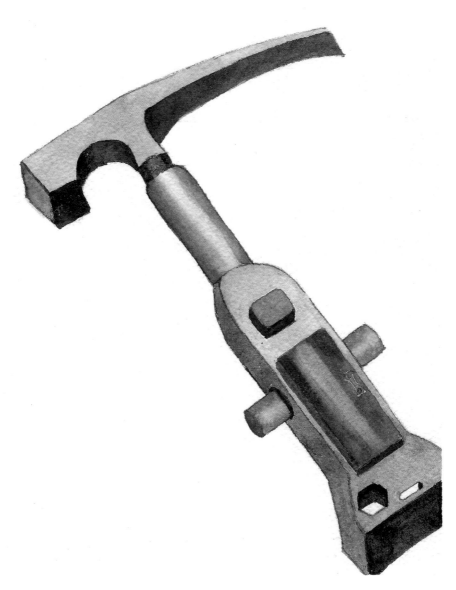

Plate 5   Prospecting on Venus's Surface

Plate 6   Surfing Saturn's Rings

Plate 7  Sailing the Titanian Sea

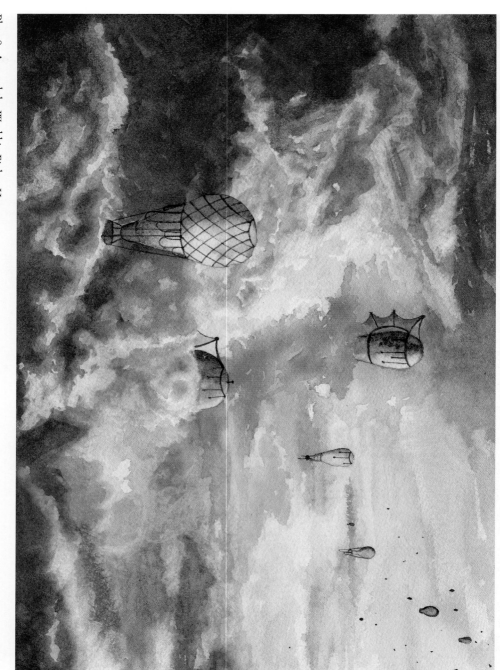

Plate 8  Around the World in Eighty Hours

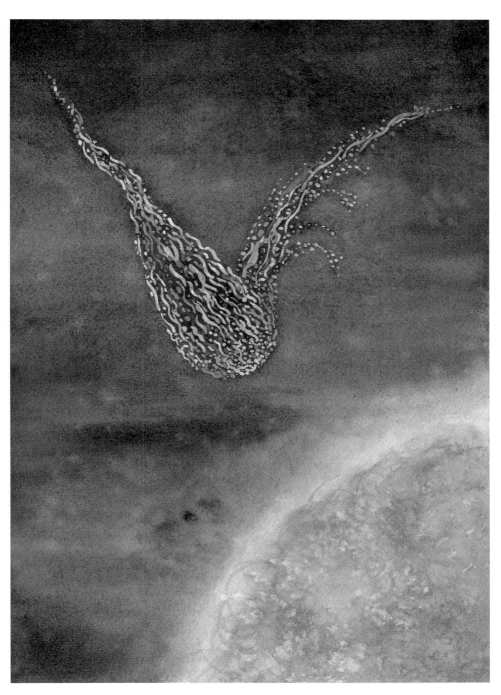

Plate 9   A Year Shipwrecked on a Comet

Plate 10   Of Roses and Baobabs on Bennu

Plate 11 Skydiving In Jupiter's Vast Atmosphere

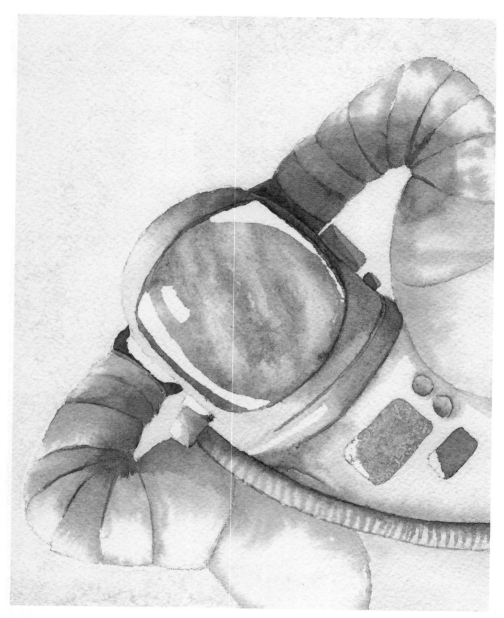

Plate 12   Watching the Martian Clouds Scoot By

Plate 13 Spelunking on Hyperion

Plate 14   Caffeinated Flight on Titan

Plate 15   Pluto, the Edge of the Map

Plate 16 We Are All Explorers

you are close enough that you feel yourself to be a well-protected part of the group—enfolded on all sides, as well as above and below, with glowing envelopes of gas. It's almost as if you're flying in a strange flock of soaring birds. Or perhaps that you're in a forest of glowing baubles.

You remember, once, in the forests of New Zealand, walking through a high tree canopy late at night. The rope gangways were perched a hundred feet off the ground, but you were constantly surrounded by enormous lanterns. Almost nothing is the same between that experience and this one. And yet almost everything feels the same. Strange, the parallels our minds invent to try to explain our perceptions.

Gradually, the Earth begins to set in the east, letting you know that the dawn is starting to approach. The safety airships allow the formation to break and scatter to make it easier to bring everyone in for a safe landing. Taking their cue, the balloons vent their envelopes and intentionally begin to lose altitude, dropping out of the high-speed air.

As dawn grows across the cloud deck below, you can start to make out a few tiny specks in the distance. From experience, you know it's a group of hotels. They look much the same as the ones you left 80 hours ago, but you know these are different. That last group of hotels has been making progress across Venus's daylit side and are, by now, thousands of kilometers away. This group of sky cities have placed themselves here purely to serve as the end point of your journey. Before the light of dawn becomes too great, they welcome the flotilla with a spectacular display of fireworks.

As you descend, smaller, more powerful dirigibles ascend from the hotel to meet you. It will be the task of a couple of hours to capture each balloon and to bring it safely into its mooring. But once you get down to the level of the hotels you'll be moving at the same speed and can afford to wait for your tugboat of a dirigible.

On the radio, the other balloonists are already celebrating—no trouble to tell the first timers from the more experienced. Once you get into the hotel, there will be a spectacular party: colorful banners and (yes) helium balloons in every shade, elated handshakes, hugs from your friends and others you know less well, and who could forget the traditional champagne that has celebrated the reunion of balloonist and solid ground since the eighteenth century?

You know some balloonists for which the round-the-world journey is nothing more than an excuse, a prologue for this party.

But for you, the journey has always been the point: that feeling of float-ing in alien skies, with the best view in the solar system. To do it all in a vehicle where the only thing you can control is your buoyancy. For you, the flight becomes about embracing that journey, your powerlessness in the face of nature. But with the confidence to know, come what may, that you'll get through it.

As the dirigible approaches to bring you into safe harbor, your thoughts turn to next year and the chance to make this journey all over again.

★★

In chapter 5, we took a good look at the surface of Venus, describing its extremely intense characteristics (high temperatures and pressures), and dis-cussed the possibility that Venus didn't always look the way it does today. Well, here we are, back again at Earth's sibling planet, Venus. Though this time, instead of the hellish reality of the prospector searching through fool's gold, possibly on a fool's errand, we have ascended to the relative comfort of the mid-atmosphere.

CLOUD CITIES!

As you climb higher in a planetary atmosphere, the pressure drops. This is because, as we described in chapter 5, pressure is caused by the weight of the atmosphere above you. The higher you go, the less atmosphere there is above you, and therefore the pressure goes down. The temperature also typically drops as you rise from the surface in a thick atmosphere like the one Venus possesses. So, the higher you get above the surface of Venus, the lower the temperature and pressure. Surprisingly, between about 50 and 60 km above the surface, there is a Earth-like layer, where the temperatures can hover near room temperature with pressures not terribly different from living in the mountains on Earth.[1]

This means that you could go ballooning on Venus wearing not much more than a sensible coat and an oxygen mask. That's right, no spacesuit required! This has led some to speculate that a sort of airborne life could exist here (but don't forget our ALH84001 mantra!). The catch is that if you go lower or higher you'll enter a volume that is not conducive to human life, so you would need to live among the clouds. Luckily, ballooning on

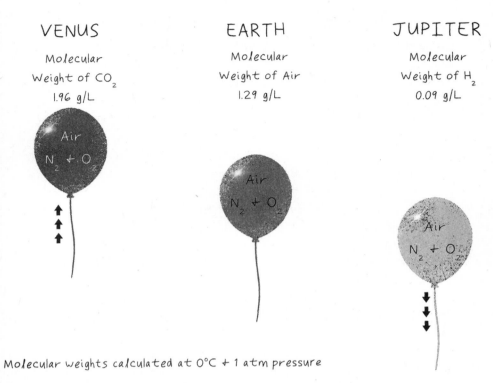

VENUS

Molecular
Weight of CO$_2$
1.96 g/L

EARTH

Molecular
Weight of Air
1.29 g/L

JUPITER

Molecular
Weight of H$_2$
0.09 g/L

Air
N$_2$ + O$_2$

Air
N$_2$ + O$_2$

Air
N$_2$ + O$_2$

Molecular weights calculated at 0°C + 1 atm pressure

Figure 8.1
A balloon filled with breathable Earth air that would be stationary at home would naturally rise on Venus and would sink like a stone on Jupiter because of the relative densities of the gases that make up each atmosphere.

Venus is relatively easy because of that nearly pure carbon dioxide atmosphere. $CO_2$ is one of the densest molecules that regularly forms a major component of planetary atmospheres. Each molecule of $CO_2$ weighs 22 times more than what a single molecule of molecular hydrogen ($H_2$) weighs, which is the primary constituent of gas giant planets, and each $CO_2$ molecule weighs about 1.5 times more than a single molecule of molecular nitrogen ($N_2$), which is the primary constituent of the atmospheres of Titan or the Earth. See figure 8.1.

That means that almost anything that isn't $CO_2$ can be used to fill up a Venusian balloon, even the air we like to breathe as humans, which is a nitrogen-oxygen mix. If we chose this route, humans could live right inside the balloon envelope itself. Gone are the cramped quarters of most human-rated spaceships. Instead, you would need wide-open, vaulted, and airy spaces to provide lift. This means that an effective structure floating in the atmosphere on Venus would likely look like a giant balloon itself or, as in the story, a series of balloons connected to resemble a cloud. A literal cloud city!

Unfortunately, there are issues with floating at 50–60 km above the surface of Venus, regardless of the temperate weather you may find there. Starting at about 50 km above the surface and extending to an altitude of 70 km, there is a very thick layer of clouds made mostly of sulfuric acid ($H_2SO_4$). There are no full breaks in this cloud deck, which never entirely dissipates, meaning that it's impossible to orbit above Venus and take pictures of the surface using visible light.[2] This is different from Earth, where there are always clear views through to the ground.[3]

Clouds are created when gas in the atmosphere condenses into liquids or solids. On Earth, our atmospheric temperature and pressure favor the condensation of water. That means that as you ascend into our atmosphere you will invariably find places where the temperature and pressure require water to turn into a liquid from its gaseous states (i.e., condense).

On Venus, the combination of temperature and pressure is wildly different from Earth's, but that doesn't mean condensation is impossible; quite the contrary. It just means that the materials that can condense out are not water. Incredibly, the conditions on Venus allow sulfuric acid ($H_2SO_4$) to condense out, creating entire clouds of acid. On Earth, we have sulfuric acid in our atmosphere too, both naturally and put there by various industrial processes. Some of this sulfuric acid can dissolve into the water droplets that are condensing in our atmosphere to create rain that is more acidic than normal, a.k.a. acid rain.

Surprisingly, Venus's clouds of sulfuric acid do produce rain (if they become saturated enough). This means that, literally, acid can rain from the sky. However, as the acid falls, it gets closer to the surface where the temperature is higher. Before it hits the ground, the temperature gets too high, and the acid evaporates again. This is a phenomenon we sometimes see on Earth: when precipitation evaporates before it hits the surface, it is called virga.

Since Venus is constantly enveloped by sulfuric acid clouds, preventing any view of the surface, the only way to "see" the ground is by sending radio light through the clouds, letting it bounce off whatever geological structures it hits, and then measuring the return. If the radio light takes a longer interval to return, then it had to go farther, and you're mapping ground that is at a lower altitude; if it takes a shorter interval to return, the ground is higher. This approach can be very accurate, down to a few centimeters, creating highly detailed maps using orbiting spacecraft. The first successful mission to do this for Venus was NASA's Magellan probe from the early 1990s, which used radar to make the first full map of the Venusian surface.[4]

## ROTATION AND SUPER ROTATION

While it's fun to think about how similar this portion of the atmosphere is to Earth's, it's the super rotation of Venus's atmosphere that lends itself so nicely to this daydream.

A planet's spin is the amount of time it takes to rotate on its north-south axis once. Of course, as we learned in chapter 4, things get complicated when you realize you can measure this relative to different things. But either way you measure it, one thing is for sure: Venus spins slowly. The sidereal day on Venus is 243 Earth days—that's longer than the time it takes Venus to orbit around the Sun, which is 225 Earth days. A solar day on Venus, however, is about 115 Earth days. Remember, a sidereal day is the time it takes the planet to rotate, where a solar day is the time it takes for the Sun to rise, set, and rise again, from the perspective of someone standing on the surface. Even more interesting, Venus is spinning in the opposite direction compared to the other major objects in our solar system. This leads to an odd effect: the Sun rises in the west and sets in the east.

At the surface of Venus, winds are rather mild, clocking in around 5 km/h. This is comparable to the rate of rotation of the planet itself, meaning that if the hot air ballooner was trying to circumnavigate the entire globe of Venus with the balloon near the surface, it would take something like a couple hundred days. The reason the winds are so calm at the surface is because there is no major change temperature from place to place. Winds are driven by energy differences. With such a thick atmosphere (both in the Venusian and Titanian atmospheres) all the Sun's energy gets deposited at high altitude. This prevents the lower atmosphere from having any major energy differentials, allowing it to come to a calm equilibrium.

But at the mid- to high altitudes in Venus's atmosphere, the winds are moving much faster, as much as 60 times faster than the speed of the planet's rotation itself (or hundreds of kilometers per hour). A hot-air ballooner bobbing along in this high-speed wind would be able to circle the entire planet in just a few days![5]

It truly is an atmosphere of weirdness on Venus. Atmospheres and their characteristics vary considerably and can even be temporary, as in the case of comets when they get close to the Sun. Indeed, we're coming face to face with a comet in the next chapter.

# A YEAR SHIPWRECKED ON A COMET

"Now you've done it," you say out loud as frustration fills your thoughts. "Of all things, shipwrecked on a comet!"

How did you end up here? Well, that's a simple story. As a comet miner, you have become used to a life of excitement. You have made a career of swooping in fast, hot on the heels of any new discovery. You harvest the water, the nitrogen, and anything exotic on these icy puffballs and then you get out quickly and sell your cargo to the highest bidder back on the planets. Then it's on to the next comet.

But this time you got a bit too greedy.

The old timers warned you not to touch anything inside Mars's orbit, especially not long-period comets full of fast-evaporating ices. Once these celestial bodies start to sublimate, they become unstable—throwing off boulders with no warning, breaking up into city-sized pieces as the sun's heat finds a pocket of dangerously unstable hypervolatile ices. But you didn't listen and hooked up to a comet barely outside the orbit of the Earth, one with a beautiful puffy atmosphere and tail already forming.

When that building-sized fragment came at you, it's a wonder your ship didn't crack open like an egg! Instead, it was just your engine that was crushed and mangled. Your ship is going to be here for a very, very long time.

At least you'll be okay—you're not in any danger of dying. The safety regulations for miners dictate that you have a couple of years of dried food and enough water and oxygen to make it through. If you wanted more, it would be easy to make it. There's enough water ice right outside your window to fill a few million swimming pools and solar power enough to break it all apart into its constituent oxygen and hydrogen atoms. You could even make your own rocket fuel from that stuff. But without an engine, even a full tank is useless to you.

But not to worry, it gets worse!

As soon as you realized your predicament, you radioed back home for rescue, as humiliating as that was. Unfortunately, given that your comet is already too close to the Sun for comfort and has proven to be somewhat unstable, the company doesn't intend to run a rescue operation until your comet finishes its close pass to the Sun and gets far enough away on the other side to cool and for its activity to subside. Plus, given that you're stranded on a special type of comet called a sungrazer—a comet that comes very close to the Sun, speeding up enormously as it approaches the heart of the solar system—it would certainly be a very costly rescue.

As a result, you're on your own for the next year, until celestial mechanics aligns planets and flight paths for you again. For someone used to a fast-paced life, the prospect of this enforced slow time seems suffocating.

At least you know you won't be bored as you wait. No, no—you have months of back-breaking work ahead of you. In theory, it's possible to survive a solar passage on a sufficiently large sungrazer, like your current unwanted celestial companion. But to do so, you must pay close attention to how the comet is constructed, how it moves and rotates. That's the only way to stay protected from such a close pass to the thermonuclear fire that is our Sun, tethered to nothing more than a rapidly melting snowball.

Lesser sungrazers typically give up the ghost—evaporating away to nothing on their close encounter. Telescopes see them go behind the Sun, but they don't see those comets come out the other side. Had you been mining one of those, your fate would have been sealed by the tyranny of gravity. But anything much larger than a few kilometers in size, like this one, can make the passage, shrinking greatly in the process—bulk traded, with gusto, for survival.

The next day, you start your work, resigned to your fate. What a way to start January. You unpack your equipment to scan the interior of the comet. That means trucking the sensors out by hand and laying them over the comet's surface. By placing ground-penetrating radar and seismometers, you can get an idea of the density distribution within. Unlike planets, whose early processes cause them to separate out into onion-like layers of different density, comets have been frozen since their formation in an amalgam of different, randomly collected bits and pieces frost-welded together. That means that you'll need to find out, ahead of time, whether there are other pockets of hypervolatile ices that will lead to more fragmentation or other weaknesses in the comet that could endanger your survival.

Once you know this object, inside and out, it's not just a matter of finding the best depression or cave in which to hide. This comet, like most comets, rotates and will start rotating faster and more erratically as it evaporates more in some places and less in others. These evaporation sources, known as "jets," act like little rocket thrusters. Luckily, your spaceship has enough computing power to know just how to shift the most easily evaporated materials around on the surface to balance something this big. But it's you who must do the shifting by hand, with a shovel.

As a result, your days become filled with routine. And then more routine. And slowly, ever so slowly, something inside of you seems to shift along with the piles of snow.

At first, you can't put your finger on what that something is. Maybe it's something in the air? As the days pass and then the weeks, the angry god in the sky that is the Sun becomes larger and larger. Meanwhile, the environment of the cometary surface becomes stranger and stranger. The super-thin atmosphere around the comet, called the coma, becomes thicker and larger and even whitish in color. If it looks like a lightbulb illuminating an impossible number of dust motes swirling around the nucleus, that's because this is exactly what is happening. That milky glow lends a strange flat light to the foreshortened landscape.

Your instruments now tell you that this region of gas and dust is nearly as large as the Sun itself. At the edges of the coma, these dust particles interact with the momentum of the sunlight photons and slow down, being drawn backward to form the comet's characteristic tail. Meanwhile, the sun's radiation takes apart the evaporating hydrogen and oxygen of the water molecules, ionizing them and allowing them to be swept away by the solar wind—charged particles following magnetic field lines. Because the magnetic field lines spiral away from the Sun rather than flowing directly away, this ion tail can stick off the coma and nucleus of the comet at apparently strange angles, appearing as a strange blue appendage to those observing through a telescope from afar.

Both features are only barely visible to you, even on the side of the comet away from the Sun. The dust tail is like a slightly brighter white apostrophe in the "sky" and the ion tail is like a faint blue afterimage that moves around relative to the dust tail as the magnetic field twists and bends.

But no matter where you are on the comet you can no longer see the stars, which surprises you. On most planets, the height of the atmosphere is small compared with the diameter of the planet. So when night falls at a particular location on the surface, it follows quickly even in the high atmosphere—though you will often observe illuminated clouds for a few minutes after sunset even on Earth. But the comet is different. Here, the atmosphere is so much larger than the nucleus itself that even the "night" side of the comet is bathed in reflected light from the floating dust particles that would be "over the horizon" on Earth. That constant light reflected into your eyes ensures that they cannot sensitize themselves to the dark conditions needed for stargazing.

But this dream-like cast to everything, as if you were walking in a cloud, is not the only thing changing around you. More and more, you find yourself paying close attention to details in the landscape. You notice strange formations carved into ice, reminding you that a comet like this one spends most of the time in its millions of years–long orbit much farther from the Sun than Pluto.

Indeed, the vigor with which this comet is evaporating suggests that this may be its first trip close to the inner solar system since it formed billions of years ago. Perhaps it had a sedate life out there in the Oort cloud—the swarm of comets that surrounds our solar system—before a close encounter with another object changed its trajectory just enough to send it hurtling sunward.

Strange, you think, that something left untouched for so long is fated to become nothing more than raw material for civilization. But at the same time, collecting the water and bringing it home means that this material can participate in the cosmic dance of life. Is being consumed by humans a better or a worse fate than evaporating near the Sun and becoming a part of the interplanetary medium? You can't answer that question, and you realize that the origins and fate of the comets you had been harvesting is not something to which you'd previously given much thought.

But now, either way, the comet seems to be becoming a living and breathing thing. The practical implications of this activity force philosophical considerations from your mind. This particular bit of frosty sand or icy rock may have lain here for the age of the solar system until it had the misfortune to find itself under your shovel, but its movement will not be in vain as you move the material to where your computer tells you it needs to go to keep the comet from spinning.

That evening, after a long, hot shower (there are advantages to being marooned next to an almost inexhaustible source of water), you strap yourself into a chair near your computer. Lately, you've been passing the evening hours writing letters and recording messages to old friends and family from home. There's a thrill to reconnecting with people in your past. You hadn't realized with many of them that you had been out of touch for so long. You're shocked at how they've grown older and realize that you too must look a bit more aged to them.

They tell you stories of partners, of children; of a simple life full of birthdays, graduations, and quiet nights watching the sunset with loved ones. It all has an attraction you had never even contemplated. Though you're unconvinced that you want the same life that your friends have, it disturbs you that out here, in the wide universe full of possibilities to chase, you might have forgotten about the existence of other less heralded, but no less meaningful paths.

It's strange to connect through a screen like this. But at least you can easily tell them where you are. Your friends and family need only look up at the sky after dark to see one of the century's great comets. The press has even renamed it after you!

Wanting to share a little more of your little piece of the heavens, you've begun to set up cameras here and there. Scientists back on Earth use the pictures that those cameras send back to analyze how the coma is changing and to get new views of the Sun. No one has ever tried to physically alter a comet in this way before, and at least you can have the satisfaction that a few people will earn PhDs from your sweat equity. Sometimes, you take photos of particularly interesting formations to send home or fly some of your robots deep into the coma to get interesting angles, or to keep tabs on the Sun. Some of these videos become viral sensations across the solar system, and you are baffled by the attention.

Then, one day, it gets hard to see right in front of your face—the exhalations of the comet surround you, and as materials shift beneath your feet the vibrations of the comet's groans are transmitted into your suit through your boots and the metal of the ship itself. It won't be long now before perihelion passage. Unable to continue to work the surface, you retreat into your ship. You've long since moved it to the best spot on the comet. Having done what you can, now you do what you must and prepare to wait. The computers assure you that everything will be fine. But it's not in the human psyche to fully trust any mechanism. You certainly have never been a person of faith. But you don't have any choice in the matter—you have been given a harsh reminder that your fate is controlled by forces greater than yourself. A coronal mass ejection or poorly timed prominence from the Sun is all it would take to reduce your hard work to literal ashes.

Deprived of mindless labor to fill your days, you brood. Lately, the messages from home have started to show the passage of time not just since you lost touch, but now since you reconnected. Life moves on, back on the planets. Births, anniversaries. And then, just before you pass behind the Sun and temporarily out of contact with the human race, you receive news of the death of a close friend. It hits you like a hammer blow, like an impacting asteroid. And there's no work to which you can escape, nowhere to run to as you examine the smoldering crater in your life. Like the space left by a lost tooth, the gap pulls at your attention, insistent.

To distract yourself, you try to block out the light with curtains, but somehow the impossibly close Sun forces photons into the ship through every crack. Cornered, your cabin becomes a crucible. You find yourself considering deep questions: How deliberate was your choice to become a miner? Does it hold any meaning for you or have you just been following an ephemeral high these past few years? What was the cost of the achievement and acclaim you've garnered within your professional community? Wide-eyed, alone in the blinding brightness of your wrecked ship, you realize you never did take the time for any of these introspections.

Later, you look back at the tapes of your cameras and robots. They are a sight to behold. The interactions between the Sun and the coma conspire to create strange patterns of light that hold you rapt. It's only several hours later that the larger truth sinks in—that the comet, though diminished, is still there. That you are still alive.

Back in contact on the other side, everyone from home wants a piece of you. How does it feel that no one has been as close to the Sun in history as you? Or to hold the solar system's speed record, since sungrazing comets near perihelion are the fastest things in the solar system? You are dumbfounded but recover enough poise to give them the answers they think they want.

Idly, you think that both of those records would have meant something to you once. You might have sacrificed to achieve them. But somehow, they don't feel quite so important to you now.

Instead, you are seized by a fresh energy. It will still be some weeks before the comet stabilizes enough for you to get out onto its surface, but already you have plans to explore this fresh new world uncovered by the Sun. You message eagerly with scientists and debate the best way to examine that surface as deliberately as possible. What are the best traverses? Where can the best samples be acquired? You've got the better part of six months until rescue, so there's no time to waste!

Over that time, your ship becomes a laboratory, and you discover a zeal for scientific investigation you never knew you possessed. But you temper that work ethic. You know too well of your propensity to become engrossed in a task to the exclusion of all else. Leavening your activities is a new commitment to keeping in touch with loved ones and a certain Zen calm out on the comet's surface. You find it's important to appreciate the moment as fully as possible.

When the company comes to pick you up in December, they have good news for you. They sold your footage and your story and have decided to provide you with a handsome finder's fee. They're even going to salvage your ship and a part of the comet to sell to a museum as a piece of space exploration history. The rest of the comet will, of course, be harvested— can't let those volatiles go to waste out in the far reaches of the solar system.

It seems some things never change. But some things do, and you are not the same person who was shipwrecked here a year ago. The funds mean that you don't have to mine comets for a living anymore. But even if you could, that career belonged to someone else whom you barely recognize. Instead, you cast your gaze outward, thinking about places where you could make yourself useful.

You've heard that Titan is lovely this time of year.

★★

Comets are one of the most amazing phenomena you can observe in our nighttime (and sometimes daytime) sky. Their imagery appears in stories, songs, paintings, drawings, tapestries, pottery, cave walls, and any other place humans could think to describe them.[1] And who could blame us? Have you ever *seen* a comet? They're incredible. A bright yet diffuse object with a long tail streaming off into the distance. They appear in our sky and slowly move through the constellations over a few weeks or months, giving us somehow both ample and fleeting time to observe and ponder their existence, only to disappear and (apparently) never return. Humanity has a long and storied history with comets, and we encourage the reader to take some time to investigate both our cultural and political reactions to them, as well as our evolving understanding of their nature and origin, culminating with the famous Sir Isaac Newton and Sir Edmund Halley. Unfortunately, we are unable to do that here. Instead, let's look more at what comets are physically, and what they tell us about the solar system.

## WHAT IS A COMET?

A comet is a small solar system object, maybe a few to tens of kilometers wide (like the size of a major city), made of a mixture of rock and ice. When we say "ice" we mean all manner of ices, including water, carbon dioxide, ammonia, and other substances. Comets are leftover material from the formation of the solar system, meaning that what comprises them has never been part of a large planet or moon (the same is true for asteroids, as we will see in chapter 10).

There are (probably) millions of these icy objects in our solar system, most of which hang out in the Kuiper Belt or the Oort Cloud, never coming close enough to the Sun to become a comet as we classically understand one. But what is the Kuiper Belt? What is the Oort Cloud?

The Kuiper Belt is donut-shaped torus of icy objects gravitationally bound to the Sun that exists beyond Neptune, stretching from about 30–50 au, or maybe a factor of two or three farther.[2] The Oort Cloud is a spherical *shell* of icy objects also bound to the Sun that exists well beyond the Kuiper Belt, stretching from something like 500 au to . . . well, we don't know how far. Maybe halfway to the next star, or about 100,000 au distant.[3] For the most part, the millions of icy objects that occupy these regions stay where they are. They have all manner of orbits ranging in size, eccentricity, and inclination to the plane of the solar system, but the vast majority of them don't get closer to the Sun than Neptune's orbit.

Because orbital speeds are so slow out in the Kuiper belt and Oort cloud, even gentle encounters with other objects can drastically change the orbit of these icy bodies. As a result, some of these Kuiper Belt objects (KBOs) or Oort Cloud objects (OCOs) can suddenly find themselves on extremely elliptical orbits, stretched out like pulling a rubber band between two fingers. At their farthest, they would be out in the Kuiper Belt or Oort Cloud, but their hugely elliptical orbit brings them hurtling toward the inner solar system, picking up orbital velocity, like falling down a hill toward the Sun. It's these objects that become the comets that we see in our sky, that we talk about in stories, and that a shipwrecked miner landed on in our daydream.

After a comet's orbit has been disrupted out in the Kuiper Belt or Oort Cloud, and they come screaming into the inner solar system, sometimes, they have small but influential encounters with planets. Jupiter, for example, may subtly nudge the orbits of comets and slowly, over eons, regularize their orbits into smaller ellipses with shorter periods. These short period comets come back around in our sky time and again. For example, Halley's Comet, which was possibly once part of the Kuiper Belt, fell towards the inner solar system and has been regularized to have an orbit of about 76 years around the Sun. Long period comets have had no such change encounters with larger objects in our solar system, or maybe it's their first time passing through at all, and thus often boast orbits with very large periods. Hale-Bopp, for example, takes approximately 2,500 years to orbit the Sun once.[4]

### A TALE OF TWO TAILS

Any images you find of prominent comets like Hale-Bopp or NEOWISE clearly show two tails emanating from the comet: one is faint and blueish, the other is brighter and mostly white. Why do comets have two tails? It's because there are two different types of stuff that get ejected from a comet as it is slowly evaporated away.

When one of these KBOs or OCOs approaches the Sun, ices that make up their structure turn into gas. The gas builds up around the icy object, which we now call the nucleus of the comet, forming a very large temporary "atmosphere," called the coma. For some comets, the coma can be as large as the volume of the Sun.[5]

At the outer edge of the coma, there is nowhere for molecules to hide from the intense, high-energy radiation coming from the Sun, particularly in the ultraviolet. Being directly irradiated by the Sun can cause some of the molecules to break into two fragments, each of which will carry an electric charge. We call this process ionization, and the fragments are known as ions.

The funny thing about ions is, since they are charged, they are forced to respond to magnetic fields. This electromagnetic force can be much more powerful than the gravity holding the molecule to the comet, so these ions

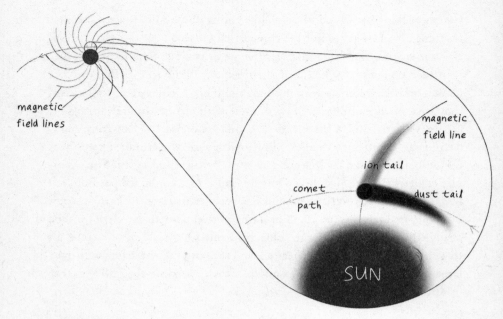

Figure 9.1
Detailed view looking down on the north pole of the Sun. Comets have two tails.
Charged gas particles called ions are affected by magnetic forces and follow the Sun's
spiraling magnetic field lines. Because the particles are so small, they interact most
strongly with short-wavelength light and appear blue. Dust particles emitted from the
comet slow down and fall away from the comet, leaving a bright white tail.

at the edge of coma promptly fly away from the comet, following the spiral
field lines that emanate from the Sun. This is depicted in figure 9.1. Since
the molecules are so tiny, smaller than the wavelength of light from the
Sun, they tend to scatter blue light best—just like our atmosphere does.
This creates a thin, and sometimes hard to see, blue tail emanating from the
comet, made of ionized coma molecules, formerly part of the comet's ices.
We call this the ion tail.

Now, while the gas was sublimating from the nucleus, it liberated dust
that had been held in place by solid ice. Since the comet itself is so small, it
has very little gravity, so the dust begins to float and disperses throughout
the coma. At the outer edge, these dust particles too come face to face with
the intense radiation from the Sun.

The dust is too big to break apart into ions, but that doesn't mean the
solar radiation has no effect. Instead, there is an interaction between the dust

and the light we call the Poynting-Robertson effect. This process begins to slow the velocity of the dust particles by a tiny amount, just enough that the dust particles begin to fall behind the comet, stretching out into a curved pathway in its wake. These dust particles are larger than the visible wavelengths of light and act a bit like reflectors, showing no preference for the wavelength of light that they scatter toward our eyes. And so, we see a large, mostly white tail stretching large distances away from the comet in our skies. This is often the much larger and more obvious tail we all know and love.

### SUNGRAZING AT PERIHELION

As discussed in chapter 1 and in much more detail in chapter 4, all orbits in the solar system are ellipses; some are just more elliptical than others. For example, the Earth's orbit around the Sun, is close to a circle, with a low eccentricity of 0.0167. A Sun-grazing comet's orbit around the Sun is extremely elliptical, with some having orbital ellipticities measuring 0.99. Thus, every object orbiting the Sun will have a closest approach and a farthest approach. The point in the orbit where the object is making the closest approach to the Sun is called the *perihelion* (Latin for "near sun"), and the farthest distance is called the *aphelion* (Latin for "away from sun"). The Earth's perihelion occurs in early January, with an Earth-Sun distance of about 0.98 au (147.1 million km), and the aphelion occurs in early July, at about 1.02 au (152.1 million km). That's right, over a six-month period from January to July the Earth gets about 5 million km farther from the Sun.[6] The average of Earth's perihelion and aphelion distance is about 149.6 million km and is the formal definition of the astronomical unit.

Looking over that last paragraph again, you might notice that Earth's perihelion occurs in January, which, if you're a resident of the Northern Hemisphere, might be a bit puzzling. If we're closest to the Sun, shouldn't it be hotter? Yet January is one of the coldest months of the year. This is because our seasons are dictated by the tilt of our north-south axis with respect to the Sun. While we do change our distance from the Sun, the difference of 5 million km doesn't have that large of an effect, but our tilted axis does.[7]

The comets that enter the inner solar system start way out in the Kuiper Belt or Oort Cloud, with aphelion distances of anywhere from 50, 100, or 1,000 au. At that far distance, the Sun's gravity is weak, and the comets are moving slowly. As they move through their orbits toward perihelion, they literally fall toward the Sun, picking up orbital velocity the closer they get. In the case of sungrazers, their perihelion distances are about 0.001 au. With simple numbers like this, we can do some velocity calculations to show how much acceleration occurs. According to our shipwrecked miner, this particular sungrazer is probably from the Oort Cloud, so let's assume that its aphelion distance is about 1,000 au. At that distance, the comet's orbital velocity is a languid 1,000 km/h or so. However, after falling 1,000 au toward the Sun and swinging around at just 0.001 au, the comet would accelerate to ludicrous speed of 1.5 million km/h! According to the Guinness World Book of Records, the fastest any human has ever traveled was achieved by the Apollo 10 astronauts Tom Stafford, John Young, and Eugene Cernan, who reached a speed of 39,937.7 km/h on their trip to the Moon and back.[8] Hitching a ride aboard a sun-grazing comet would definitely smash that record.

## THE COMET-ASTEROID SPECTRUM

The typical definition of a comet is rocky-icy object, whereas an asteroid is solely a rocky object. Since comets contain both rock and ice, and when they get close to the Sun they lose some of that ice, there is a spectrum of different kinds of bodies that exist in between the ideal of a rocky asteroid and an icy comet. In reality, all small bodies are somewhere in between, with some looking more rocky and others looking more icy.

A dusty, boulder-strewn comet can evaporate all the ice on its surface and form a protective lag of fluffy dust. This can insulate and protect the ice below from any further evaporation. To all the world, this object now looks just like a dusty, rocky asteroid! It even stays completely inert when passing close to the Sun with no coma and no tails forming.

At the other end, processes within presumptive asteroids can dredge up water from inside from time to time or can have their internal stores of volatile ices exposed in collisions with other asteroids. A few years before its encounter with the Dawn spacecraft, the largest asteroid in the belt, a dwarf planet called Ceres, was observed to have a very faint comet-like water vapor tail that hinted at the evaporation of water.

Because the forces involved are small with most comets, it's not typically dangerous to approach a nucleus. There could be some larger chunks of dust and ice that have detached and are floating around nearby, but without much gravity it's a quiescent place. Even so-called jets on comets are just places where more sublimation is happening than usual. If you were tethered to a comet near a jet while it was "erupting," you might not even know that anything was happening. Indeed, the European Space Agency achieved just such a feat, when they approached and landed on the surface of Comet 67P/Churyumov–Gerasimenko leading up to and during its perihelion. When the Philae lander touched down and began collecting data, it found a quiet and dusty almost snowy environment, generated by the sublimation of ices all around it.[9]

However, all of the forces that sublimate a comet and create a tail are magnified on sungrazers, the comets that, like the one in our story, pass extremely close to the Sun. Here the solar radiation and the subsequent heating is extreme. Pockets of very volatile ices that exist below the surface can be heated, causing them to vaporize and explode, completely disrupting a comet. It's not unusual for those observing the Sun through telescopes to see a sungrazer disappear behind one side of the Sun and then to see many pieces coming out on the other side.[10] Or, like Icarus after the wax in his wings melted, for nothing remotely resembling a distinct object at all to emerge—the comet instead being reduced to a diffuse cloud of gas and dust by its passage.

It truly was fortunate that our protagonist was able to modify their comet to avoid such a fate!

In this chapter, we investigated comets; in the next we look at a similar type of object: the asteroid. Both comets and asteroids have much to teach us about our solar system and have already been the destination of many missions. For example, in 2016 the OSIRIS-REx mission launched from Earth to the asteroid Bennu. It rendezvoused with Bennu in 2018, orbited there for three years, collected a sample, and returned that sample to Earth in September 2023. Bennu is the topic of our next chapter.

# 10

## OF ROSES AND BAOBABS ON BENNU

All these years later, you can still clearly remember the moment when you first encountered Antoine de Saint-Exupéry's *Le Petit Prince*. You couldn't have been more than six years old. It started as a bedtime story, but you were so entranced by the tale of the prince, his travels, and his asteroid that you would surreptitiously take the book out during the day when your parents were busy. Then, at night, under the covers with a flashlight, you read ahead, marveling at the watercolor designs and the prince's clear-eyed but innocent wonder at everything he encountered. It's even possible that, in some small way, the vividness of that story might explain a small part of why you eventually chose to study asteroids as a scientist when you were grown.

Once you began studying, the prince's fairytale world quickly fell away. Small asteroids like the one he inhabited can't support an atmosphere—the gas molecules we breathe are always in motion and there's just too little gravity on such tiny objects to keep those molecules from flying away into space. Such a small asteroid would also cool far too quickly to allow anything resembling the terrestrial volcanism described in the book to take place. So, no volcanoes, dormant or not. Indeed, in many ways, you wouldn't expect a small asteroid to even have a solid surface on which you could stand. Many of these bodies are simply piles of smaller pebbles and boulders held together by exceptionally weak gravity. "Rubble piles," you and your colleagues call them.

Don't blame Saint-Exupéry here: there was no way for him to know about these discoveries when he was writing. Yet these realizations ultimately didn't temper your enthusiasm for asteroids. For every deletion of fairytale magic, larger and more fascinating puzzles revealed themselves, each more significant for being grounded in the real world. Far from being mere lifeless piles of rock, asteroids preserve information about the early solar system and the processes that were important billions of years ago.

Alone, each one is a time capsule of ancient chemicals and minerals. In the aggregate, their motions and groupings tell us of the historical roaming of our solar system's giant planets.

And so, the kernel of interest from the novel has stayed with you. It is a small point, shining brightly at the center of your studies. And, by any measure, you've had a good career of it to this point. You've studied some interesting problems, nurtured some excellent young scientists, and made some insightful contributions to the field that have impressed your colleagues. In fact, you even have an asteroid named after you, as many scientists in your field do.

But recently, you've been feeling that a change is needed. You've found that interior light dimming, perhaps clouded by the events and the years you have accumulated. Maybe there's a way to recapture the magic, if you only got out from behind your desk, to see these places with your own eyes, rather than simply looking at the data sent back from robotic spacecraft and telescopes. It seems a little crazy, but what you really want to do is some gardening. If you could only grow a rose in the broken-up stone regolith of an asteroid, maybe you could reconnect to something you've lost. After the year that you've had, certainly no one would begrudge you some time away to indulge that sort of eccentric pursuit.

Ahh, but they do.

"Why are you throwing your career away?" they ask. "You were set, on the tenure track!"

"But isn't science about following your curiosity," you respond, "supporting one another to find and create knowledge from which everyone can benefit?"

"Absolutely not!" they say. "It's about publishing as many papers as possible in top journals and bringing in the grants. It's about pushing your students to the edge where they can make the discoveries that you will own. It's about winning prizes to prove to everyone else just how great your intellect is."

After enough of these conversations, you have discovered there's no advantage to engaging. These are articles of faith to many of them—it makes you wonder about what lies at the core of their motivation. Better to travel onward and leave those scientists to their own devices as if they were content beavering away on their own isolated asteroids, alone in the vastness of space.

But not everyone feels the same. In your quest, you are deeply moved by colleagues who offered their excitement and their help. A biologist who set you up with some UV-tolerant seeds. A physicist who designed and built for you a bell jar to house your plant. An engineer who provided a system to break down the local rocks, distilling water while leaching out unwanted salts and other chemicals that could be detrimental to your garden. These will all complement your standard kit for surviving in space.

With these gifts, you head off on your journey. You've decided to set up on Bennu, a small but well-studied near-Earth object. Bennu is quite a bit larger than the prince's asteroid, at just over half a kilometer in diameter, but the larger size will make it a bit easier to get around and for all your gear to stay put. Even so, the light gravity, more than 100,000 times less than that on the Earth, means that you'll still need to be careful: pushing off from the surface with more than 8 cm/s will send you on an escape trajectory, never to return.

There are other benefits. The carbonaceous nature of this object should give you the best raw materials with which to set up your garden. Because it's an Apollo asteroid, it orbits between Earth and Venus, so there's plenty of light. You also know from the samples returned by the OSIRIS-REx mission that there are particles on the surface of the right size to feed into your equipment for processing. Finally, though Bennu is a bit of a backwater today, its status as a potentially hazardous asteroid means that you can hitch a ride with rangers who periodically check up on these objects and make certain that nothing has changed that could set them on a collision course with Earth.

After a few weeks of travel you arrive, waving goodbye to the departing ranger's spacecraft and assessing your dark and rocky new home for the next few months. You have tethered your little habitation module, or "hab" for short, to the boulder Benben. You picked this 30-meter-tall solid protrusion as your base camp because it's something to which you can reliably attach your gear. No matter how you move about in your hab, the block has enough inertia that you know you can't make it move and accidentally float away from the asteroid.

The next few days are all about getting self-sufficient and running ropes along the surface of your redoubt and down to the smaller rocky bits below. You don't have enough safety line to circle all of Bennu, so you practice with the compressed jet backpack that will save your life if you accidentally shove off in your explorations with more than the gentlest push while you are exploring away from Benben. The hardest part to master is managing the panic that makes you want to apply too much thrust too quickly, rather than just giving the physics the time it needs to work in this low-gravity world.

Once you know the hills and dales of Bennu well, it's time to get to work. Like any garden, the biggest choice is where to site your plot. In the end you decide on a part of the asteroid that the OSIRIS-REx mission called "Sandpiper." You decide this is your spot because of the hints of hydrated minerals in the exposed regolith. Perhaps material from under the surface will be even more abundant in water—the one molecule in short supply here on Bennu, so close to the Sun, and such a critical component for any garden.

So it happens that one day, perhaps a week after you arrived, that you package up the bell jar (with seeds already inside) and the regolith processor into a large net and carry them across the surface—though "carry" is the wrong word here. While you effectively crawl across the surface on all fours, you make use of the low gravity to heft the package into the sky where it makes a long and lazy nine-and-a-half-minute parabolic flight at the end of your ten-meter tether. It's a handy configuration because it keeps all your equipment free of obstacles and debris. But if ever there were a passer-by, it must look a ridiculous sight—as if you had a steel balloon over your head as you traveled.

Finally, you arrive at Sandpiper and do your best to glue your equipment to the largest rocks you can find. You burrow into the asteroid with your hands to expose hydrated regolith "soil" and then vacuum that soil into the hopper of the processor (digging with a shovel would give the rocky debris enough speed for it to leave Bennu forever). With your helmet pressed to the equipment so you can hear its inner workings, the regolith processor hums to life. A few hours later, a little light comes on, indicating that your soil is ready. You hook up the output hose to the bell jar, which, at the flick of a switch, begins to fill with dark material like ground-up chocolate cookies.

The soil swirls and mixes in eddies, like a malevolent snow globe or a wizard's crystal ball. Every so often, you get a view of the lighter seeds before they disappear back into the tempest. Some of the soil seems to be clumping together—a good sign! Clumping usually means water. Under Bennu's microscopic gravity, you know it will take the material some time to settle down through the atmosphere inside the jar. So, for today, you content yourself with a sample of the soil from the processor and head on back to your home away from home.

When you come back a couple of days later, you are rewarded by a nice layer of soil at the bottom of the bell jar and—perhaps your eyes deceive you?—you think there's a bit of green from a seed up against the side of the jar! You can barely contain your excitement. Inspecting the processor, you also see that it has managed to distill a small bit of water from the regolith you put in earlier. You take the detachable water container, connect the nozzle to the bell jar, and are rewarded with a pleasing mist.

Over the next week, you return daily to water your sprouting roses. Eventually, three of the seeds germinate, and it is a joy to watch the little leaves get bigger and higher each day.

But then two of your sprouts stop growing. What could be going on? You consider several theories: Could it be something in the soil or in the water, or the extra ultraviolet light? It can't be the latter, as UV wavelengths would be filtered out by the specialized glass of the jar. In desperation, you add in a bit of terrestrial fertilizer—roses need more nitrogen than you can typically get on asteroids. But even with this intervention, your sprouts do not recover. Over several days they turn brown, wither, and collapse—all but one, which remains right in the middle of the jar.

Unfortunately, you don't have the equipment to do an exhaustive test for the problem; you are on an asteroid, after all. Instead, you're stuck with logic. If there's a problem with the soil, there's nothing you can do. You simply don't have another source. All you can do is change the water source and the lighting.

So, you pick up the bell jar and bring it back with you to Benben and set it right outside your hab. You seem to recall that some kinds of roses like to have some shade. Maybe that applies to space varietals as well. Next, you switch out the water for the rose with some of your own water from the hab. That earthly water has been recycled several times, passing through your body in myriad ways. Perhaps, you think, some good can come from what was once your tears.

Somehow, with the changes, your rose thrives. Just a day before you're scheduled to depart, it flowers into a single perfect red bloom. You stare at it for what seems like hours, awoken from your reverie only when you and the rose fall into deep shadow.

The next day, you break down the hab and load it aboard the ranger's shuttle. The pilot is impressed by your gardening and encourages you to bring the flower back with you. "You could write a paper on that plant!" But instead, you decide to leave it in place. Though the rose won't survive much longer, it just doesn't feel right taking it with you. Perhaps there's some value to leaving something here that is so very much of this place, but also contains more than a little bit of yourself: your hopes for the future and maybe even some of the sorrows of your past.

Your last view of the rose is from the departing shuttle. Even though you must be too far away to make it out, you would swear you can still see that bright red blossom against the gray and black of Bennu's surface.

## WHAT EXACTLY IS A RUBBLE PILE?

Of all the daydreams in our chronicle of the imagination, the scenario of the gardener on Bennu is the only time we imagine the low-gravity, unprocessed environment of one of the solar system's many millions of asteroids. An asteroid is another type of small solar system object, ranging in size from the very small, as little as a few meters in size, to the very large, hundreds of kilometers in size. Most of these asteroids hang out between Mars and Jupiter, a region we call the asteroid belt.[1] There we find the largest true asteroid (meaning, one that is not a dwarf planet), Vesta, which is about 530 km in size.[2] Asteroids are certainly not confined to the belt. We find them all over the solar system, as exemplified by the setting of this chapter: Bennu's perihelion is at 0.9 au, which is inside Earth's orbit, and its aphelion is at 1.3 au, outside Earth's orbit. This means Earth and Bennu share the same region of space (or, as the gardener puts it, Bennu is a near-Earth object).

Bennu is only 500 meters in diameter, a very small place.[3] If we think back to what affects surface gravity (size and mass; see chapter 1), Bennu ends up having such a low surface gravity that even standing on it wouldn't be very easy. A quick calculation indicates the surface gravity is about 0.000006 $g$, or about 6 micro-$g$. With such a low gravitational strength, even very small pushes from your feet would have you achieve escape velocity and leave the surface of Bennu forever. This low-gravity environment makes it difficult (albeit comical) to get around for our gardener. Since gravity is so low, it makes Bennu much less like the small, solid world of *Le*

*Petit Prince* and much more like a pile of rocks of various sizes held together loosely by gravity.

Imagine a dump truck with a load full of random rocks. The dump truck lifts its bed and lets all the random rocks pile into a heap on the ground. The rocks are all piled together, but you can easily imagine that inside the pile are many open spaces where the rigid rocks of various sizes cannot squish completely together. The amount of mass at the top of this imaginary pile of rocks is not strong enough to break the rocks and squish them all together (we discussed this in chapter 2). If an object gets big enough, if the imaginary dump truck just keeps dumping load after load of rocks, eventually, there would be enough mass to squish and break the rocks down to get rid of the empty spaces.[4]

From surface to center, asteroids the size of Bennu simply do not have enough mass to crush rocks; thus, Bennu is like the pile of rocks from the dump truck: a big pile of rubble mashed together. In fact, on Bennu the gravity is so weak and the rocks are so loosely packed together that the asteroid can randomly eject rocks and boulders through natural processes like resettling. Even further, NASA has described its surface to be like a plastic ball pit. Imagine jumping into a ball pit. You would sink right in! We know this because Bennu was the destination of a mission called OSIRIS-REx. This mission flew to Bennu, orbited it, descended to the surface, used its robotic arm to retrieve a sample of rocks, and flew back to Earth. While there, OSIRIS-REx observed the asteroid randomly ejecting rocks.[5]

GARDENING WITH A TIME CAPSULE

The problem with growing a rose on Bennu is not the gravity; it's the lack of the important materials needed for life. If we remember back to figure 0.2, our solar system, and all of the materials in it, began as a giant cloud of gas and dust that we refer to as the solar nebula. Through the forces of gravity, the conservation of energy and momentum, and many millions of years, we went from a nebulous cloud to a mostly flat solar system with a star, planets, moons, asteroids, and comets. You might think, then, that the chemicals and materials we find on Earth were present in that original solar nebula. Following that logic, we should be able to dig up a sample of Earth, measure all of the chemicals in it, and thereby know what the solar nebula was made of before the Earth started forming.

Unfortunately, this leaves out a crucial piece of the puzzle, and that is the geological and biological processing. The Earth is so big and massive that, through a variety of processes such as differentiation, volcanoes, plate tectonics, convection, decay, and biological metabolism, the chemicals here have changed throughout time. Most rocks on Earth are new rocks, formed relatively recently.

In order to know what the solar nebula was made of, we have to examine objects that have been relatively unprocessed since the beginning of the solar system. Objects that are too small for plate tectonics, or volcanic eruptions, or melting of rocks. Objects like Bennu. Since it is a rubble pile, there's not enough gravity here to make anything geophysical happen. This is why we often refer to Bennu and other smaller asteroids like it as "time capsules" of the early solar system. The rocks that make up Bennu formed early on in our history, are some of the first rocks to exist here, and have been relatively unchanged since the origin of our solar system. They are a snapshot of our early days. By sampling Bennu and other asteroids, we can get true measurements of the material available when the solar system was forming. Figure 10.1 shows a pie chart demonstrating the different types of materials in asteroids similar to Bennu.

The most important chemicals that are needed for life on Earth are carbon, hydrogen, oxygen, nitrogen, phosphorus, and sulfur (a.k.a. CHONPS). It's easy to see from the pie chart in the figure that, while some of these were found in the solar nebula in large amounts (e.g., C, H, O), nitrogen and phosphorus were in very short supply. If you tried to grind up some Bennu rocks into a sand in which you hoped to plant flowers, you'd be missing some key ingredients.

The simple conclusion here is, if you put a few rose seeds in random Earth soil and added some water and sunlight, you would grow a rose with relative ease. If you added a rose seed to regolith made from Bennu rocks and added water and sunlight, the plant could not grow, because there is not enough nitrogen and phosphorus and what exists of these two elements are often in non-bioavailable forms.

So why has Earth been able to make up the difference? Why does Earth have a whole bunch of readily available nitrogen, phosphorus, and sulfur? There are many geochemical and biological processes here that act to move these materials around, for example, the carbon, nitrogen, and phosphorus

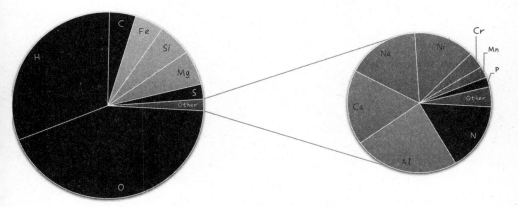

Figure 10.1
Relative composition of elements found in the oldest solar system materials (such as on asteroids like Bennu). Primitive objects like Bennu have compositions similar to the earliest solar system materials. For those materials, such as CI chondrite meteorites, the balance of different elements is well known, as represented by the pie chart. The elements required for life are highlighted in black, and, just like on Earth, the least abundant is phosphorus, in this case nearly 1,000 times less abundant than oxygen. Data source: Katharina Lodders, "Solar System Abundances and Condensation Temperatures of the Elements," *Astrophysical Journal* 591, no. 2 (2003): 1220–1247, https://doi.org/10.1086/375492.

cycles. These cycles have made it possible for life to make use of, and produce more of, the materials it needs.[6]

Thus, soil needed for gardening is much more than crushed-up rocks.

Our next destination is the largest planet in the solar system: Jupiter. If you added up all the mass of every other object in the solar system (other than the Sun), together, they would still not have as much mass as Jupiter. It is truly *massive*. Gas giant planets are a wholly different world compared with the rocky planets. Let's look at what it would be like to visit and enter the clouds of Jupiter.

## SKYDIVING IN JUPITER'S VAST ATMOSPHERE

As you ready for your skydiving flight, you gaze out the window at Jupiter spinning below. The station passing over the planet's equator brings you to within 5,000 km of the cloud tops near the 1-atm pressure level of Jupiter's atmosphere. You ponder, briefly, why the 1-atm level should be altitude zero for gas giants. It seems very Earth-centric, given that it's the sea-level pressure on humanity's home world. But in a bottomless atmosphere, you suppose the mark needs to be set somewhere, just to help navigation, if nothing else. It's also a great level to find clouds here on Jupiter.

The station has a polar orbit, so that the scientific instruments and the staff of researchers who call this place home can observe every latitude of the planet every three hours. The orbit stays in the same orientation relative to the sun—great for keeping solar panels powered—and Jupiter's rotation takes care of the rest, turning underneath and revealing all longitudes as the station passes overhead. Thus, over time, the full complexity of the atmosphere is unveiled.

That rotation is a complex dance, taking just under ten hours on average. But, like the trade winds and jet streams from home, Jupiter's atmosphere is a windy place full of atmospheric zones that form visible bands moving in alternating directions, with terrible speeds of up to 500 km/h, or nearly 10 percent of the rotational speed of the planet. These violent winds are powered from within, by the leftover energy of Jupiter's formation, billions of years ago. To this day, the planet is warmer and shines brighter in the infrared than you would expect, given how much energy it absorbs from the Sun.

You arrived here 45 minutes ago so that you could get in the show's most dramatic transition: the station moving over the north pole, the vivid colors that swirl together but do not blend. Above 60°N is the polar maelstrom where storms dwarfing terrestrial hurricanes play bumper cars on a colossal scale. Meanwhile, in the mid-latitudes and equatorial atmospheric bands, more orderly clouds endlessly circle in a slower dance that can take decades or even centuries to unfold.

And now, your quarry begins to rotate into view—the famous Great Red Spot, a storm wider than the entire Earth that has churned away for at least 400 years. This storm is different than most. Not a cyclone but an anti-cyclone, rotating in the opposite direction than hurricanes do on the Earth. The color comes from the access to the layers that lie below, with material dredged up by the storm's dynamic atmospheric engine.

An announcement from your pilot tells you that it's time to launch—so you climb into the spaceplane and fasten your harness. It's a relatively easy task to which you have become accustomed here in zero gravity. You are the only skydiver on this flight, so you nod to the pilot and are off.

The pilot points the engine opposite the direction of the station's motion and taps on the thrusters to reduce the speed of the spaceplane from its 150,000 km/h motion relative to the planet. Jupiter's gravity responds as you fall even more firmly into the grip of the most massive planet in the solar system, and you appear to fall down and in front of the station in the nonintuitive dance of orbital mechanics.

At first, the trip is uneventful. Gradually, you get closer to the features in the Jovian atmosphere and their true scale becomes apparent. Storms the size of planets are made up of swirling clouds that gain more texture as you approach. White pinpricks that looked like strange little bumps from the station reveal themselves to be herds of thunderstorm anvil clouds, each 50 km tall, boiling up out of unseen cauldrons below and crackling with lightning. At this altitude, each detail is gone in a flash as you pass overhead while covering more than 40 km every second.

At 450 km above the 1-atm level, the spaceplane starts to feel the very slight resistance of the atmosphere, and by 200 km of altitude you formally cross Jupiter's Kármán line and the spaceplane begins to be supported more by lift from the wings instead of its orbital motion.

The entry into the Jovian atmosphere is the most challenging transition in the solar system. Intuitively, you know that the spaceplane needs to slow down, dramatically, from its orbital speed to become an aircraft. But you have no idea how that is achieved by this vehicle, nor how it is protecting you from being crushed. You recall that, centuries ago, the Galileo probe felt a force 228 times the force of gravity on the Earth during its atmospheric entry. So you are left to wonder at the white-purple hydrogen plasma surrounding the craft as it slows, an eerie dancing light visible in every window that gives a strange illumination to the otherwise dark cabin interior.

Less than three minutes later, it's over. The plasma clears and the aircraft is flying straight and level in the Jovian atmosphere. You know this, intuitively, from the sensation of your weight having more than doubled. If everything went right, you should be about 15 km (about 49,000 feet) above the 1-atm level, and indeed, below you is a cloudscape unlike anything you've seen before.

Red- and blue-tinted swirls intertwine with white but never seem to mix completely. Some speculate that the "chromophores" that give the clouds their color might be made up of exotic compounds of sulfur or phosphorus, but no one really knows. This close, the clouds have a remarkable mottled texture to them.

But before you give this skyscape a more detailed appraisal, you look upward. Just like on the Earth, the sky here is a clear and deep crystal blue —violet from the scattering of the sun's light by the molecules in the atmosphere. It reminds you of a full Earth moon on a mountain top—but at Jupiter, the Sun is 27 times fainter than as seen from Earth.

At this altitude, there is effectively no longer a horizon to see. On Earth, you would be traveling high in the stratosphere, just a few thousand feet above the cruising altitude of a long-haul transoceanic jet. From that perch, you might strain to make out an object 400 km away. But here, the more subtle curvature due to Jupiter's massive size means that the same altitude puts the horizon nearly 1,500 km away. So, the sky doesn't so much meet the cloud tops as the two appear to dissolve into one another in the great distance, with the clouds bluing into the darker blue-violet of the sky like impossibly distant mountains.

But as you look more closely at that hazy vanishing point, you realize that the blue is slowly turning red ahead of you. This color extends across your forward view and gradually begins to loom, forming a great red wall ahead. It is the edge of the Great Red Spot, whose clouds loom eight kilometers above the surrounding cloudscape, rising halfway up to your cruising altitude. Your destination.

Below, the clouds seem to form a giant river in the sky, carrying you forward. At the red wall, the river parts, flowing cleanly around. As you make the transition to flying over the Red Spot itself, the journey begins to become turbulent, with your craft being gently tossed this way and that. Indeed, sound waves created by the roiling of the Great Red Spot are constantly heating the atmosphere above.

You're interested to know more about what's happening down below in the depths of the clouds. That would be a challenge for the aircraft you're on—at this altitude its airspeed is still 2,500 km/h. Even kilometers into the cloud deck, you'd still need more than 1,000 km/h of speed to stay aloft. If you're going to take a closer look, you're going to have to go on your own.

Jupiter may be a tough place for flying, but it's a great place for falling!

With a word to the pilot and the push of a button, you and your skydiving rig are deployed. You experience a moment of weightlessness as you fall away from the aircraft. At first, the wind whines loudly against your faceplate, but in short order, you slow down to terminal velocity. With just a pressure suit and a parachute pack, that's still about 2,000 km/h, but this time it is directed downward instead of along the cloud tops. That's faster than the fastest free-falls on Earth by stratospheric jumpers by a good margin, and is about 60 percent of the speed of sound here on Jupiter.

For a second, you revel in the 360-degree view of the largest sky in the solar system, temporarily lost in an endless atmosphere that appears to be all your own. But only for a second—at this speed, you'll arrive at the cloud tops in less than a quarter of a minute, so it's time to deploy your parachute.

First, a smaller drogue extends to slow you part of the way, and then the larger, 16-meter-diameter parachute follows. This main chute is truly massive—nearly the same size as the parachutes used to slow the Curiosity rover in its descent to Mars. This drops your speed to a stately 140 km/h or so.

At this point, everything has become much quieter, almost leisurely. You glance at your altitude reading and see that you still have about three minutes before you hit the cloud tops. It's an important time to check your kit to see that all your instruments and sampling canisters are ready to go. You try out your communications link to the aircraft far above—you can see your pilot circling your drop zone and you give them a thumbs-up, even though there is no way they can see the gesture at this distance. Given that it's −120°C outside, you're thankful for the suit. It will get colder still just before you hit the clouds, bottoming out at a bit colder than −150°C before the pressure and the temperature start to increase, the deeper you travel. Indeed, your suit has more in common with deep-sea diving couture than with anything astronauts are known to favor.

As you are looking upward, you pass through the first fleecy wisps of cloud, made of ammonia ice crystals. Instead of the hexagonal forms made by water, ammonia crystals form cubic and octahedral forms, which give unearthly optical effects. You see two haloes around the Sun, a little wider apart than what you would see on Earth, as well as a bright arc above the Sun. To the sides, where you are used to seeing two water ice sundogs, you instead see four boomerang-shaped rainbow features, like chandelier crystals at the ends of imaginary diagonal lines drawn through the sun's disk. Somehow it seems fitting that Jupiter can put on the best show of atmospheric optics in the solar system.

You want to ponder these features longer. But Jupiter's gravity keeps pulling you downward, and within a few more seconds, you are fully enveloped by the topmost cloud layer and the Sun is lost except for a diffuse, milky-white glow.

It's time to get to work.

Luckily, the deeper you go into Jupiter's atmosphere, the slower you travel, which gives you some time to complete your task. Embedded in the top-most ammonia ice clouds, you are falling now at only 90 km/h, and by the time you reach your pickup point, 140 km below, you will have slowed to only a touch over 20 km/h. All told, this gives you a few hours to take an unbelievably close-up look at Jupiter's atmosphere.

Rapidly, as you descend, enough ice particles go by overhead that the light from the Sun fades away completely and you are left in total darkness. You marvel for an instant at being in such a place, so far from home, so alien. Your universe contracts inward to just the suit around you and a few inches in front of your faceplate where your readouts illuminate the white snow in garish colors.

Then, suddenly, your world expands once more. Spotlights turn on, illuminating a bubble of snow, perhaps ten meters across. This is by design and happens automatically. In darkness, time plays tricks on the human mind, and a minute can feel like an hour or vice versa. Better to leave safety to the computer.

In some ways the illuminated bubble is even stranger. If you didn't know you were deep within the largest atmosphere in the solar system, you might imagine you were in the artificial environment of a soundstage. You can just make out the ghostly underside of the parachute far above as the snow is pushed gently this way and that by turbulence. But by and large you move through the blizzard, downward. In some ways it's like looking out the window of an aircraft flying through a cold cloud, the particles illuminated by the landing lights or in the flash of the rear wingtip strobes.

Except here instead of watching the particles fly backward over the wing, they seem to be falling upward. You know, intuitively, that these particles are all falling downward too—just as all cloud particles do on every planet. They're simply not falling as quickly as you are. Eventually, they too will fall out of the ammonia-saturated layer and will evaporate to vapor, eventually mixing upward to form new ammonia snow particles.

As you descend, your suit does the background work of measuring the local weather—temperature, pressure, and the relative balance between different gases in the atmosphere. Measurements of water are taken that give you relative humidity, still very low up here in the ammonia cloud layer. Because of all the ice particles churning around you, you don't need your equipment to know that ammonia is fully saturated here. You wonder to yourself whether that means that the "ammonidity" is at 100 percent. Chuckling silently to yourself, you glance at the readouts for the suit and see that oxygen is still flowing well inside your protective bubble and the pressure inside is still just normal for sea level on Earth—no nitrogen narcosis in this suit!

You take the time to gather some samples for analysis later. While the simple things—how many ice particles you are passing through, their shapes and sizes—can be recorded directly by your equipment, you have deeper questions that can't be answered by the limited instrumentation available to you, hanging on the end of a tether below a parachute. Perhaps the samples you take will help when analyzed with scientific instruments that are far too bulky to bring to Jupiter's atmosphere directly.

For instance, did any of the chemistry taking place in these clouds ever produce something we might consider alive? Traveling up and down on eddies, in endless vertical circulation, could there be tiny organisms that eke out some kind of unique living here? Perhaps, once you get these canisters back to the lab, there will be surprises yet to be uncovered.

It is an interesting question to ponder as you continue your descent. Down and down you travel, eventually coming out of the ammonia cloud layer and into a patch of clear air. You feel yourself falling away from this surface, and for a few minutes, your spotlights reveal nothing, all their luminous power gobbled up by the scale of this open layer in between the clouds. You lose all references and could be standing still for all you know. The only companions here in this interregnum are the occasional dust particles of meteoritic smoke, or the long-chain organic molecules called tholins created by UV chemistry high above.

Enveloped by this blank space, your imagination runs wild. In your mind's eye, you imagine that you can see the ammonia clouds above, like some massive ceiling, stretching on to that fuzzy horizon 1,500 km away. Kilometers below, the waiting ammonium hydrosulfide clouds form a terrible floor hurtling upward toward you and stretching on in a parallel plane toward the vanishing point. In some places, eddies bring the floor and the ceiling together in columns of monstrous scale, like bulbous stalactites. Through it all, the feeling of deep interiority, of being inside of a planet, is overwhelming, like some unimaginably vast, vaulted Moria-like subterranean complex.

You are saved from this reverie by your arrival at the ammonium hydrosulfide layer. A fuzzy brightness below you appears to rise upward until you pass through the floor in a flash and return to your bright snow globe existence lit by spotlights. Again, you see the blizzard of ice crystals, but these are subtly different from those above. The additional complexity of the chemistry here with the addition of sulfur into the mix opens even more possibilities for coloration. It's hard to tell with the illumination you have—the mind tries to turn any bright reflection into "white" when it lacks for another reference—but you are certain you catch glimpses of more complex hues as you fall through this layer.

Another gap and then the water ice layer. This one is by far the thickest of all the layers, and the size of your illuminated bubble contracts inward. After hydrogen, oxygen is the most common element in our solar system, making up most of the atoms in the Earth when incorporated into rock, or the bulk of the icy moons when combined with hydrogen. On the gas giants, it becomes the bottom-most of the clouds.

These clouds are very much like cold clouds on Earth. The pressure at this depth, approaching ten times the sea-level pressure on Earth, is not enough to shake them from their hexagonal structure. Nor are the temperatures so different here. What is different is the distance that they can fall and the density of the air, allowing each of the crystals time to grow to an unusual size and to stick to one another. Water is the universal solvent, and perhaps, somewhere far below, it exists in a fluid form. But up here everything is still frozen—vapor and solids circulating in an endless dance.

Eventually, you come out from under the water ice layer as the last cloud deck falls away above you. The Great Red Spot continues, deeper and deeper, its roots continuing onward for another few hundred kilometers below you, 50–100 times deeper below the cloud tops than the average depth of the seafloor below the waves on the Earth's oceans.

It was intentional that you picked the Great Red Spot for this journey. If you had followed the Galileo probe's trajectory, which explored a clear "hot spot," by the time you arrived so deeply into Jupiter's atmosphere, the temperature would be well above boiling, approaching suit-melting levels for which even the best heat exchangers would have trouble compensating. But the active churning of the spot works to minimize differences in temperature between the layers. Therefore, by the time you arrive here at your pickup point, the temperature has only risen to a balmy −70°C or so.

Here in the emptiness under the clouds, your journey is nearly done; time to signal the aircraft to collect you and your samples. Down here, the pressure has risen to several dozen times the sea-level pressure on Earth, not unlike what you might find under a thousand feet of water back home. But the density and the viscosity of this material is far less, allowing you to move your arms and legs with ease as you wait.

Off to the right you see your ride approach, beginning as a point of light that drops away and rises again toward you from below on a gentle blue flame. Your fall speed is now sufficiently slow, at about 20 km/h, that latching on to the spaceplane is a breeze. You give the pilot a thumbs-up before releasing your tether and parachute and stepping inside. You watch the white mass fall away from the spaceplane. The parachute will continue to descend, eventually melting, merging, and finally dispersing into the unfathomable deep of Jupiter's atmosphere to become a part of the planet. This is the fate of almost all hardware deployed on Jupiter—to become one with the largest planet in the solar system. You wonder: How many molecules that were once part of the Galileo probe did you encounter on the way down? Not many, but not zero either. Silent, microscopic companions from home, now part of something larger.

Your last thought, before you pressurize the cabin and remove your helmet, is why one would bother to skydive on Jupiter at all, when the spaceplane can hover like this. Though there might be some contamination from the engines, perhaps it would have made more sense to do things that way. But it certainly wouldn't have been as much fun.

★★

Up until this point, all our daydreaming in the solar system has been in terrestrial places (save the rings of Saturn). They had firm ground, made of rocks or ice, and we can imagine or at least compare them with Earth much more easily. Now we are entering the world of the Jovians.[1]

Jupiter is one of the four gas giant planets, all of which occupy the outer part of the solar system, beyond the asteroid belt but within the Kuiper Belt. Starting from closest to the Sun, the Jovians are Jupiter, Saturn, Uranus, and Neptune. In complete contrast to the terrestrials, the Jovian planets are composed primarily of hydrogen and helium (very little, if any, rocks), are absolutely gigantic (5–10 times wider than Earth or more), have a range of ring systems (Saturn's is the most extensive, of course), and they each have a very large amount of natural satellites (Saturn has the most, at 146 as of June 2023).

### ON JUPITER YOU CAN SEE FOR MILES AND MILES AND MILES

When we visited the Jupiter system in chapter 3, it was to dive down into the depths of the Europan ocean. Now, back at 5.2 au from the Sun, we're exploring the depths of Jupiter itself. It's hard to fathom just how big Jupiter really is up close. The Earth's radius—the distance between the center of the planet and the surface, as measured at the equator—is 6,371 km, whereas Jupiter's radius is 11 times larger, at 71,492 km.[2] An unexpected outcome of a planet that large is that the horizon is much farther away.

The distance to the horizon scales with planetary radius. It's easiest to show this graphically. Imagine being located at the same height above both Earth and Jupiter, two planets of very different sizes, as shown in figure 11.1. From this vantage point, you could see a much farther distance on Jupiter than Earth because the line to the horizon on Jupiter (B, 1,500 km) is much longer than horizon line on the Earth (A, 400 km).

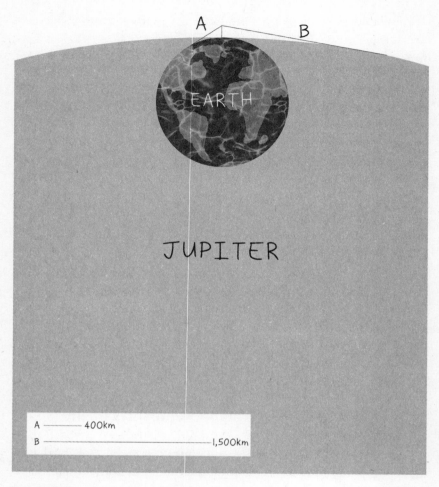

Figure 11.1
Horizon lines on planets get longer as the size of the planet increases, so you can see farther on big planets. From the same distance above both Earth and Jupiter, the distance to the horizon of Earth is line A at 400 km, whereas the distance to the horizon of Jupiter is line B at 1,500 km.

## GRAVITATIONAL GIANT

Being so large has other consequences as well. For example, this affects the strength of surface gravity. As we learned in chapter 1, the surface gravity of an object depends on both the size and mass, and we already learned that Jupiter's radius is 71,492 km. The only thing missing is mass, which is 318 Earth masses.[3] Calculating, we find that its surface gravity is 2.5 times what you would feel on the Earth. That may not be as much as you thought it would be, but you have to remember that, while Jupiter is 318 times more massive than Earth, all of that mass is spread out over a very large volume. Even still, 2.5 *g* would be hard to ignore if you were there. For instance, if you weigh 200 lbs on Earth, you would feel like you weigh 500 lbs on Jupiter.

As an aside, it is kind of funny to discuss "surface gravity" in a place that has no surface! Jupiter is a gas giant planet and is primarily made of hydrogen. As you fall into the atmosphere, you would never hit a surface, just falling forever until you reach a temperature and pressure impossible for humans to survive. Though that doesn't mean there isn't a way to feel 2.5 *g*. When the skydiver first arrives at Jupiter, they are flying in the atmosphere, which means using lift to counteract gravity. Thus, anyone sitting on that spaceplane would feel Jupiter's 2.5 *g* worth of gravity, in just the same way you feel Earth's 1 *g* worth of gravity when on a passenger jet flying at 10 km above the surface of the Earth.

The additional gravity means that the skydiver falls faster through the atmosphere. It also means that, to stay in orbit around this planet, you need move faster in the direction parallel to the surface. On Earth, to stay in orbit a few hundred kilometers above the surface, a satellite needs to travel on the order of 30,000 km/h. For Jupiter, at a similar altitude above the cloud tops, that velocity would be 150,000 km/h. Jupiter is simply pulling harder, and so you need to move faster to keep yourself from falling into the planet.[4]

This also means you need to shed a lot of velocity if you want to move from a standard orbit to actually moving with the breeze (a.k.a. flying in the atmosphere). Luckily, the most efficient way to do this is to use the atmosphere itself to slow you down, a concept known as *aerobraking*. This generates a lot of heat, and so a heat shield is needed. These are usually designed to absorb the heat and melt away as you slow down.

In practice, whether it's astronauts coming back from the International Space Station or a mission team landing a rover on Mars, there's often a combination of retro rockets, aerobraking, and parachutes that ultimately achieve the goal.

Once you have taken care of your orbital speed, falling in a planetary atmosphere is a balancing act between gravity, acting to speed up your fall, and air resistance, through a force known as drag, acting to slow you down. Gravity's pull is constant, but the faster you fall, the more drag you generate. The speed you have when both forces are equal is called your *terminal velocity*. That velocity ultimately depends on your size and shape, something you can control easily with a bigger or smaller parachute. The bigger the chute, the more drag you have and the slower your fall speed.

However, it's not always preferable to fall at a slower terminal velocity. For one thing, that means you must pack a bigger, heavier parachute. Spending more weight and volume on that parachute then means that you cannot take as many instruments with you on the ride and will therefore do less science. Additionally, you might have some timing constraints. For instance, if you're communicating via another spacecraft, it will eventually pass over the horizon and you will lose the signal. Also, your batteries may have a limited lifetime (if you are a probe) or you might have limited oxygen (if you're an astronaut). This means that it might actually be better to fall faster and see a larger section of the atmosphere. Both the Galileo[5] and Huygens[6] probes took this approach on their descents in Jupiter's and Titan's atmospheres.

## COLOR AND CLARITY OF THE JOVIAN ATMOSPHERE

How do we know what color and how clear the Jovian atmosphere would be? It turns out that these depend on reliable laws of how atmospheric particles interact with light.

By "particles" we mean a variety of things: the individual molecules of atmospheric gases (like $CO_2$ gas or $N_2$ gas), dust particles suspended in the atmosphere, or cloud particles. These particles are all different sizes. Atmospheric gases, like $CO_2$ gas or $N_2$, are the smallest on the list, measuring less than 0.1 nanometer in size (that's one 10 billionth of a meter, or 0.0000000001 meters). Dust particles have a very large range in size, but for the most part we can say they are about a micron in size (that's one millionth of a meter, or 0.000001 meters). That makes a dust particle about 10,000 times larger than a molecule of gas. On terrestrial planets, dust comes from breaking down surface rocks. However, on a gas giant like Jupiter, this material comes from photochemical processes acting on atmospheric gases (e.g., producing tholins) and from material left over from rocks that enter the atmosphere at high speed and burn up into dust, also known as meteoritic smoke. Finally, cloud particles are larger even still, usually in the tens to hundreds of microns in size, and of course, when they get big enough, we just call them rain or snow.

Light is also a very broad term that refers to electromagnetic radiation of differing energies. The entire spectrum of light, from lowest energy to highest, is radio, microwave, infrared, visible, ultraviolet, X-ray, and gamma ray.[7] We can also describe these as having different wavelengths. Radio light has the longest wavelength, gamma ray light has the shortest wavelength.

For the purpose of attempting to determine how the Jovian or any other planetary sky would be perceived by the human eye, when we refer to "light" we're mostly referring to the spectrum of light that the human eye is capable of seeing. This is known as "visible light'" or "optical light," and is basically the rainbow of colors you know and love. A classic simplification of the visible light spectrum is ROYGBIV, or red, orange, yellow, green, blue, indigo, and violet. Just as the rest of the electromagnetic spectrum is organized by wavelength, so, too, is the visible portion of it. Red light has the longest wavelength at 700 nm, and violet light has the shortest wavelength at around 400 nm. For simplicity, and from convention, scientists often refer to these two extremes as the red end and the blue end of the visible light spectrum.

There are two ways interaction between atmospheric particles and visible light can happen: absorption or scattering. Absorption is where the particles within the atmosphere can absorb the energy from certain wavelengths better than others, using that energy to heat up. We experience this effect every day when we look at objects around us. That green plant is good at absorbing all wavelengths of light except green, which it reflects. The orange bowl of fruit does the same thing but with orange light instead of green. With the right mixture of materials, a sky colored by absorption can be any color at all.

Scattering, instead, is an interaction with light in which light is not absorbed by the atmospheric particles, but instead sent off in another direction. Depending on the particle size and shape, and the wavelength of light it encounters, that effect can be strong or weak. For instance, particles that are very small (like gas molecules) tend to interact most strongly with very small wavelengths of light (like blue light). This means that they scatter blue light much better than red light. A clear and cloudless sky is blue for a reason: light from the Sun encounters what looks like empty sky, and the blue part of that sunlight is scattered in all directions, scattering many times off many gas molecules, eventually entering our eye from all directions, thereby giving the impression that the sky is colored blue.[8] This is called Rayleigh scattering, and is also the reason the wings of blue jays appear blue.

Scattered light tends to produce atmospheres that have one of three colors. Atmospheres with just gases tend to be blue in color because the gas molecules are very small compared with the wavelength of the light. Cloud particles, however, are very large compared with visible light and tend to interact with all wavelengths evenly, giving a milky white color. Dust particles, especially those of similar size to the wavelengths with which they interact, can produce any color depending on the details of the particles, but most tend to produce scattering in the yellow to red range of the spectrum (depending on their own absorption of different wavelengths).

For Jupiter, you have a clear sky made of hydrogen overlying thick clouds, so in the high atmosphere the sky should be blue, and once you get down into the clouds, you should get a milky hue. The brightness of that blue depends entirely on how much atmosphere there is between you and the Sun. This is why, when you climb a high mountain, the sky gets darker—there are fewer gas molecules above you to scatter the sun's light into sky brightness. Ultimately, when you reach orbit, there is no gas above you and the sky is black.

On Jupiter, there's enough scattering gas to produce a sky that's similar to what you would see on a high mountain on Earth when you are looking overhead. But the situation changes as you look down toward the horizon. As the amount of atmosphere that you are looking through increases, the chance of a light photon interacting with a gas molecule and getting scattered in your direction increases. Therefore, the sky gets brighter, eventually obscuring any details located far away from you and replacing them with a uniform bluish hue. You can see this effect when you look at distant mountains.

This bluing is compounded by a second property of gases called opacity in which a photon headed toward you is either absorbed by a molecule or scattered into a new direction. While absorption (converting a photon into heating up the molecules of the air) is straight-forward, opacity from scattering is a bit more complicated. If a gas is scattering light from the Sun, changing the direction of each photon and making it appear as if the sky is blue, it can do the exact same thing to light coming toward you from a distant object. That light will also get scattered. and since that light is equally scattered into all directions, most of that light will be scattered away from your eyes.

The opacity of the Earth's atmosphere on a clear day is about 0.2, and this low value is why we can see stars clearly at night through the atmosphere. However, once you hit an opacity of 1, it's more likely that any particular photon heading your way is scattered into a new direction away from you rather than successfully running the gauntlet of molecular interactions all the way to your eyes. Any object seen at this opacity starts to appear dim and fuzzy.

As a result of these two effects, Jupiter's horizon wouldn't appear sharp to an observer, but would fade gradually into the cloud tops.

PRESSURE, TEMPERATURE, AND ALTITUDE

As you move up and down in an atmosphere, there are predictable changes in temperature, pressure, and density with altitude. Typically, the temperature is high and the pressure and density are low at the edge of space—here, the molecules are most exposed to the radiation from the Sun (especially very high energy radiation) and heat up. As a result, this part of the atmosphere is called the *thermosphere.*

The pressure and density are low here due to an effect called hydrostatic equilibrium. The pressure at any level in the atmosphere depends on the gravity and how many molecules are above that point in the atmosphere. Essentially, pressure balances the weight of those molecules. Being near to the edge of space, there are few molecules above at this layer, so the pressure is low, and the density, which is a function of pressure and temperature (remember chapter 5, figure 5.1), also turns out to be low. As you descend, the density and pressure increase because there are more and more molecules overhead and the temperature decreases because each molecule is shielded by more and more molecules above from the high-energy radiation of the Sun.

Eventually, there is enough absorption of high-energy radiation that the temperature of the air stabilizes. This region of the atmosphere is called the *stratosphere*, and as we descend, the pressure and density continue to increase while the temperature stays mostly the same. This is the coldest part of the atmosphere. On Earth, the structure of this part of the atmosphere is different, due to oxygen, which splits into UV-absorbing ozone and heats up. But Jupiter does not have enough free oxygen in its atmosphere to create ozone.

Finally, in the lower part of the atmosphere, called the *troposphere*, the temperature starts to increase again. However, this temperature increase has little to do with radiation and everything to do with compression. You've likely encountered this effect in your daily life. If you've ever had the experience of pumping up a bicycle tire by hand, you may have noticed that the piston of the pump gets hot—that's heat coming from the gas that you compressed. The opposite process occurs when using a container of canned air: expansion of the gas coming out of the nozzle makes the can feel cold. Again, take a look at figure 5.1 in chapter 5.

Any air parcels that rise in the troposphere will expand and cool and any parcels that descend will compress and increase in temperature. That means that this is the ideal place for clouds to form. As the Sun shines down into this part of the atmosphere, very slight differences in absorbed energy cause motion to occur, and when a parcel of air cools enough, then water,

Figure 11.2
Temperature and pressure in Jupiter's layered atmosphere as compared with Earth's atmosphere. The black curved lines represent the temperature at some given altitude. The top of the atmosphere is the most exposed to radiation from the Sun and is the hottest, which is why it's called the thermosphere. Eventually, as you descend into the troposphere, the gas gets compressed by the weight of all the atmosphere overhead, which increases the temperature again, just like in a bicycle pump cylinder. The changing temperature means that gases condense out into liquid and solid particles to form clouds at different levels. The extra bump in the Earth's temperature profile is due to heating in the ozone layer at the boundary between the stratosphere and mesosphere. Jupiter lacks this ozone layer and therefore has no mesosphere.

## JUPITER

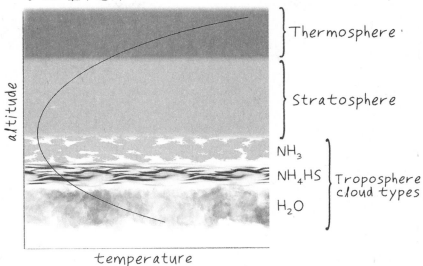

altitude

Thermosphere

Stratosphere

$NH_3$

$NH_4HS$ } Troposphere cloud types

$H_2O$

temperature

## EARTH

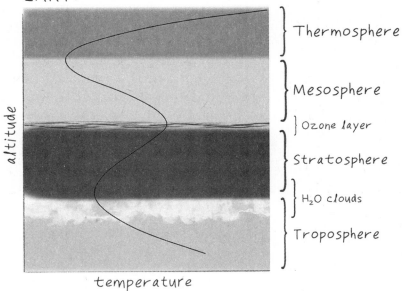

altitude

Thermosphere

Mesosphere

Ozone layer

Stratosphere

$H_2O$ clouds

Troposphere

temperature

ammonia, or ammonium hydrosulfide can condense out into liquid droplets, forming clouds. Each chemical species has a different condensation
temperature, so each forms a distinct layer in the atmosphere, giving rise to
the layer-cake cloud effect described in the skydiving scenario.[9]

Skydiving in the atmosphere of Jupiter is a rather adrenaline-inducing
activity, and perhaps after such excitement, we need to take a break and
watch the clouds scoot by. But we don't have to do that on Earth. Let's head
back to one of the more familiar places in the solar system: Mars.

# WATCHING THE MARTIAN CLOUDS SCOOT BY

You close your eyes and imagine that you are walking along a flat, dusty plain. All is silent in this place except for the crunching beneath your feet with each step. As you proceed, you sink in just a little as you break through a white crust just below the surface. Purely by accident, you strike a small rock, left undisturbed for eons.

Oops.

Startled at disturbing the quiet, you open your eyes to see the rock flying away, traveling farther, and hanging in the air longer than your earthly senses tell you is reasonable. But with just a third of Earth's gravity in this place, the logical part of your brain reassures you that this is completely normal.

So, you resume walking, continuing onward, looking for that perfect spot. Somewhere you can lie back and just stare up at the sky for a few minutes. Somewhere you can contemplate this place and the clouds scudding along overhead.

Though it might not seem like it at first, Mars has the most similar surface environment to that of the Earth of nearly any place in our solar system. An exceptionally dry, exceptionally cold environment, to be sure, but something we can recognize on human terms, even if it isn't a place where we could survive without the protection of a space suit.

Here, overnight snow is not unheard of. Ice fogs fill many low-lying areas before burning off just after dawn. Sunrise itself is dramatic: a disk one third smaller (but no less bright) than what we are used to at home throws the dusty landscape into stark relief. As the Sun rises higher, a salmon-colored dusty haze is revealed, filling the sky. Though that Sun provides less than half as much heat as we expect, when it is at its zenith the temperature of surface rocks (at least near the equator) can get as high as 30°C. That sounds like a comfortable surface against which to lie down and pause.

Through it all, the air keeps its cool. Indeed, the contrast in temperature between the surface and the cooler atmosphere can be strong enough to spawn dust devils. These features form mostly in the late morning, spinning across the landscape, picking up fine dust in their wakes. You can see a few right now, lazily churning between yourself and the distant hills. Their tops, kilometers high, are lost in the haze.

You see your spot: a sandy dune that has drifted up against a bench of rock. Over time, the thin winds of this place have pushed and piled the sand up against this obstruction. Over even longer periods of time, those same winds are taking apart the rock bench, eroding its edge further and further with each small strike of sand against stone, exposing new rock below that has been buried and preserved for billions of years.

If you had the right tools with which to ask some scientific questions, that new rock would tell you a story of an ancient lake filling the moat around a lonely mountain in the middle of a crater. It might well have been a habitable place. But was it inhabited? You cannot say; today, you've come with empty arms seeking to spend time with the Mars of today instead of a world whose time has long since passed.

Sitting down, you settle into a comfortable position and tip your head backward. Most of the time, what clouds there are on Mars are very faint. But not today. It is aphelion season, when Mars is as far from the warmth of the Sun as it can get in its circuit around our star. With that cold comes a band of clouds of water ice that stretches around the planet at the equator. Vivid streaks of white cascade across the sky in ways that remind you of the tail of a horse.

As you watch, the streaks seem almost alive. The mares' tails wisps of cloud move horizontally with the winds and seem to fall like icy waterfalls in extreme slow motion. An immense number of cloud particles, born at high altitudes, fall into drier air below. As they approach the surface, they evaporate, becoming smaller and falling slower until they disappear completely; this precipitation is called virga.

But these are not the only forms you see in the sky. From past walks you know that sometimes there are rows of clouds that appear to stretch from horizon to horizon, all traveling together like parallel streets on a map. But here there are no cars, and the streets of clouds move sideways to the direction in which the roads are laid out. In a very real sense, these clouds are showing you the surface of a layer in the atmosphere, as if you were gazing up from the seabed at the waves on the surface of the ocean.

Sometimes more complex patterns form. It's not unusual to see two separate sheets of clouds at different heights moving at different speeds and in different directions. Other times, the patterns interfere, creating a grid-like texture. One of the strangest forms consists of a cloud that seems to appear out of nowhere, then climbs over an invisible hill and disappears as it falls back to the level from which it began. It's just one more reminder that invisible atmospheric humidity and bright-white cloud particles are just two sides of the same watery coin. When the conditions are right, each can transmute into the other and back again.

On Earth, similar icy particles can bend and separate light rays to create stunning visual effects. Sundogs, arcs, circles, and haloes that create rainbow glints and bright lines in the sky are common optical features at Arctic sites and on cold, sunny days. But on Mars, it's quite rare to see these effects outside of the annual spectacular noctilucent clouds at twilight in the early spring. You didn't head out with these kinds of expectations earlier today, and the patterns of white are enough for this trip. But you can't help wonder why the conditions are so rarely quite right here on Mars. Perhaps there's something about a dust-choked sky that doesn't allow the kinds of geometric particles that play with light to form here.

After all, some days, there is so little water vapor in the atmosphere that all you can see is the swirling dust itself. Turning to look toward the Sun, you see a broad and milky brightness on the same side of the sky as the Sun. Uncountable numbers of dust particles scatter the sun's light toward you, as if the whole sky was made up of glare from a dusty windshield. That scattering is most efficient in the red, lending the sky a salmon-orange color.

The dust has its own season and its own patterns. It is thinnest in the summer and thickest in the winter, as experienced in the northern hemisphere. During that winter season, winds and dust devils in the south stir up the tiny dust particles, lifting them high into the atmosphere where they can take weeks or even months to settle out. Every few years, this process gets out of hand and enough dust is lofted that it encircles the planet—a global dust storm. The amount of dust present in these events can be so thick that it can be hard to tell just where the Sun is in the sky during the day—all you experience is a slightly brighter patch of sky instead of even the smaller-sized disk that you are used to on Mars.

You glance at the time and realize that you've been lying here pondering the clouds and the dust for hours. It's time to start moving again. With some reluctance, you fold yourself up and rise to a standing position. But instead of heading directly back the way you came, you decide to loop by an old historical site on your way.

It's a bit of a climb to where the Curiosity rover came to rest, high in the foothills of Mount Sharp. As you head up the trail, you cover in an hour parts of a course it took this spacecraft years to drive. Here and there you can see the imprints of the wheels. At some point, this will all be roped off and controlled, like any historic site. But for now, you are one of only a few people on Mars, and none would dream of disturbing the soil.

Following those tracks at a safe distance, you eventually come across the spacecraft itself. You circle around the front to get a better look and glance up at the cameras on top of the mast, above your head. For an instant, sunlight glints off the ChemCam and you have a silly thought about what the scientists would have thought if they could capture your picture right now, looking up at them. What a surprise they would have had: "Aliens on Mars!" the press would have trumpeted. It's a thought that makes you smile.

The Sun is sinking lower and lower on the horizon. Time to get back home. With a salute to the relic that accomplished so much in its time, you head down the hill. As you travel, the Sun becomes less and less bright. The light rays, passing more steeply through the atmosphere, must run the gauntlet of greater and greater quantities of dust. Often, when this happens on Earth, the disk of the Sun itself becomes a deep red color. But here, with the dust scattering away all the red light to other parts of the sky, the Sun becomes a brilliant blue-white jewel and the sky near the Sun changes color to match.

Just as you reach the habitat, you catch the last rays of the Sun over the crater rim. Phobos rises in the West, a bright and noticeably potato-shaped moon that starts streaking across the sky to set in the East. Just one more example of something hauntingly familiar in this environment, yet subtly different and exotic. As much as you want to stay outside and experience all that the Martian night can show you, you're keen to get inside. You know that the same dust that caused the spectacular sunset will also make darkness fall much faster on this planet, and the forecast called for temperatures in the −90s Celsius tonight. So, as the door opens, you slip into the airlock,

looking forward to a warm meal with good friends, no less familiar a ritual for taking place here on the dusty plains of another world beneath a
cloudy sky.

## THE COLOR OF THE MARTIAN SKY

While lost in the clouds, the daydreamer remarks on a couple of interesting
things about the color of the Martian sky. First, they notice a pale salmon-
orange color during the day, and then near the end of the day they remark
on how the Sun itself became a bluish-white color, along with the sky near
it. This is such an interesting mirror version of what happens on Earth: we
have a bright blue sky during the day, and, at sunset, the Sun turns a deep
red color along with the sky in the direction the Sun is setting.[1]

How do we know Mars's sky would look this way? There are two reasons. First, we already learned in chapter 11 all the ways that atmospheric
particles and light interact, and how these interactions create the colors we
see. Mars's atmosphere is 95 percent $CO_2$, which will preferentially scatter
blue light, just like the primarily $N_2$ atmosphere that Earth has. So why
is the daytime sky a pale salmon-orange and not blue like Earth's? This is
because the other major constituent of the Martian atmosphere is dust. A
lot of dust. Now, dust can scatter all different colors, depending on the
size of the dust grains and the amount suspended in the atmosphere. On
Mars, the dust particles happen to be a bit bigger than the wavelengths of
visible light, and so they tend to scatter a reddish hue. If you couple this
red scattering by dust with the fact that the actual amount of $CO_2$ in the
atmosphere is small, and therefore doesn't create a lot of blue light to begin
with, the dust-scattered light dominates, and you end up with a pale salmon
color during the day.[2]

However, when the Sun gets lower on the sky near sunset (or sunrise),
it means that there is more atmosphere for the sunlight to go through. The
more atmosphere the light goes through, the more dust the light encounters, and the more and more red light from the Sun gets scattered away.
Eventually, there's nothing left but bluish tinged light which the Martian
dust preferentially scatters straight forward that makes it to your eyes, and
you get a blue-white-colored Sun, as well as the sky surrounding it.

The second reason we know the Martian sky looks this way is because we
have photos of it! Mars is one of the best explored planets in our solar system,

and there are many fantastic images of some really cool phenomena. This includes the wonderful pale orange sky[3] and the oddly surreal blue sunset.[4]

Of course, as we know, if the particles are bigger than $CO_2$ or dust, such as ice crystals in clouds, then they will scatter all the wavelengths of light from the Sun. Those features are restricted to a grayscale palette. They will look white where the clouds are thin, and dark where the clouds are thick.

CLOUDS ON MARS

Mars's atmosphere is much thinner and much drier than the atmosphere on the Earth, but what little water vapor is there can still reach saturation and freeze out into water ice clouds. Unlike on Earth, though, Mars's environmental temperature and pressure range cannot produce liquid water clouds, so there's no rainfall. However, there is evidence of snowfall. We'll come back to that.

Since the conditions at the Martian surface are very similar to the terrestrial stratosphere (see chapter 11, figure 11.2), the shapes of many clouds on Mars look very much like cirrus clouds, a type of high-altitude water ice cloud that we see on the Earth. On both planets, they form wispy, painterly shapes that can look like a field of grasses or the tails of horses ("mares' tails," formally). An artist's impression of Martian mares' tails can be found in the bottom left of figure 12.1.

Even if most clouds are cirrus, there's still a wide variety of cloud shapes that arise on Mars and have been well documented by a variety of landers and rovers. While clouds can be observed at any time of year on Mars, there is a seasonally driven effect that leads to a large amount of clouds at a specific time of year. This occurs when Mars is at aphelion, farthest from the Sun, which also coincides with the northern hemisphere summer months.

Now, remember how in chapter 9 we learned that, while Earth does change its distance from the Sun, it's not a large enough change to have any meaningful effect on our temperatures? So even if Earth is at aphelion, farthest from the Sun, it's still really hot in the northern hemisphere. On Mars, the change in distance from perihelion to aphelion is large enough that it does affect what happens to both the weather and climate. Mars's perihelion distance is 1.38 au, and its aphelion distance is 1.67 au. That's a difference of 43 million km from closest to farthest, or more than eight times larger than

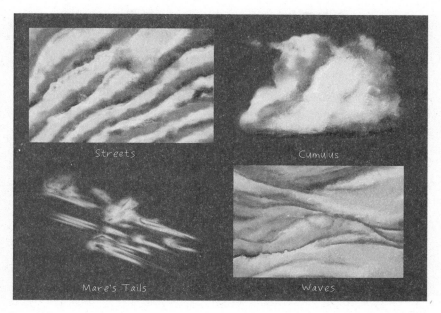

Figure 12.1
Just like on the Earth, Mars has many different shapes of clouds from organized streets of clouds to cumulus (or "heap") clouds, cirrus-like mares' tails, and even some wavy clouds that appear to climb up and over invisible hills in the sky over the course of a few minutes.

Earth's change in distance from the Sun. That means that, while it is technically summer in the northern hemisphere on Mars, the aphelion distance is so far that it actually makes things colder, leading to a higher chance of water ice crystals freezing out of the atmosphere, and more clouds being created. This phenomenon is called the Aphelion Cloud Belt, and is characterized by increased cloud formation, mostly near the equator.[5]

CUBIC SUNDOGS?

As we discuss water ice crystals floating in the atmosphere, it's tough not to think about one of our favorite atmospheric phenomena on Earth: sundogs. We use this term loosely to refer to a collection of refraction-based light effects where ice crystals hanging in the atmosphere bend sunlight traveling through them to create interesting patterns. These primarily happen on cold days, and are easiest to see when the Sun is low on the horizon.[6]

The daydreamer indicated, unfortunately, that it's rare to see these same effects on Mars. In principle, they should occur, because they are created by sunlight traveling through ice crystals, and Mars has both. A likely reason why Martian water ice clouds don't frequently produce the spectacular optical effects we get on the Earth is because there is much less water on Mars and much more dust. As a result, each ice grain that forms on a dust particle is relatively starved of the water vapor it needs to grow. This, in turn, prevents a regular crystal structure from forming.

When conditions get exceptionally cold on Mars, like high up in the atmosphere at the poles during winter months, the temperature can approach −150°C. At this temperature, not only can water vapor freeze out of the atmosphere into ice crystals, but some of the $CO_2$ can freeze out into ice crystals also. Therefore, it's possible that some of the high-altitude clouds that have been observed at the poles of Mars during winter months are actually $CO_2$ clouds. While that's interesting just on its own, it also leads us wonder what kind of sundogs they would create. The optical effects would be different from water ice clouds. Water ice crystals have a six-sided, or hexagonal, symmetry, which is why snowflakes have six arms. But $CO_2$ ice crystals have cubic symmetry, like salt crystals, which produce different displays. In figure 12.2, we illustrate what these cubic sundogs could look like (right) and compare them with what we see on Earth (left).

To cap off this discussion about plunging temperatures and ice clouds, you may ask the question, Does it actually snow on Mars? The answer is yes! Not only does it snow regular water, but it also snows $CO_2$. Can you imagine "dry ice" falling out of the air, building up on the ground, and then you making a snow angel in it? What would that feel like? Would it be different from rolling around in regular snow? These are great questions for future explorers.[7]

## WHAT WILL BECOME OF OUR ROBOTIC EXPLORERS?

The robots we send out never do come home. Eventually, their systems degrade to the point where they are no longer able to power on and they become inert relics, unable to take measurements or communicate with the Earth. If left alone, eventually all will decay and disappear, though it might take a very long time. Some might be buried by sediment, others destroyed by impacts large and small. Nevertheless, our spacecraft on other planets

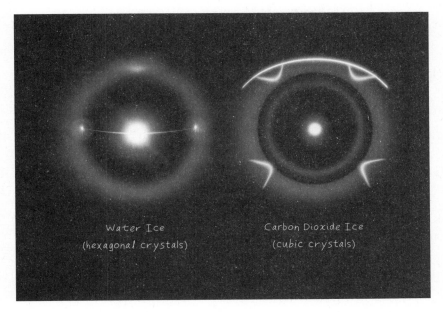

Figure 12.2
Differently shaped crystals form different optical displays. Though very few ice haloes
have been seen on Mars, theory about how cubic $CO_2$ ice crystals interact with light
(on the right) suggests very different optical displays as compared with the sundogs we
see on Earth from hexagonal water ice (on the left). The high-altitude ammonia
clouds on Jupiter also have cubic symmetry and would look similar. *Source*: Les
Cowley and Michael Schroeder, "Forecasting Martian Halos," *Sky & Telescope* (Dec.
1999): 60–64.

and especially our spacecraft traveling through interstellar space might out-
live our civilization. If anyone else ever comes looking, they may be the
only artifacts that can tell the story of who we were.[8]

However, what if we become a truly spacefaring civilization and our
robot explorers are no more than a prelude to a greater human radiation
throughout the cosmos? What do we then owe to these early objects of
exploration? Perhaps their final resting places will become museums—
preserved for future generations and revered. Or the sites surrounding
them, or any remnants of their tracks, rendered into national parks of a
sort. Certainly, the first humans to visit them will be torn—having had a
close connection with these avatars of exploration, they will want a peek at
the genuine article while hesitating to desecrate what might, for them, be
a sort of sacred site.

It's unlikely that all explorers will have such internal struggles. In the past, prizes were offered to any team that would visit and beam back photos from an Apollo landing site. There are also those who would readily turn what once was a public project into a private possession.

Let's turn our attention back to the Saturnian system now, to one of more the oddball places in our solar system. Hyperion, one of Saturn's moons, has put some solar system science to the test, and to truly find out what's going on there, we need to take a look inside.

# SPELUNKING ON HYPERION

Over the last few hours, the small moon Hyperion has gradually grown larger and larger in the window of your craft as you approach for the final descent. Saturn, its parent planet, looms large in the rear view. You swung around it just about a day ago and are now on approach to one of its infamous satellites. Hyperion, a small icy moon, is about 1.5 million km from the ringed planet, so at this distance, Saturn has shrunk to an apparent size of just a few degrees, which is smaller than your hand held at arm's length, though still much larger than either the Sun or Moon as seen from the Earth. Though you had plenty of time to enjoy Saturn vistas yesterday, today and the next few weeks are all about Hyperion.

Thinking back, it's almost unbelievable you've made it here from where you started. Back on Earth, you got into cave exploration as a kid, but you didn't have a name for it then. At that time, it was just having fun. The feelings of excitement, nervousness, and anticipation you had when climbing down into a cave, wiggling through a small opening, uncovering a new place, or just being somewhere you had never been before are still visceral. You were always the one to push the boundary, find the smallest squeeze, or the darkest room. As you grew up, you found amateur spelunking clubs that helped you push your comfort zone out of tourism and fun, and into true exploration and mapping.

If you participate in an activity long enough, you get to know people and join other expeditions, and you've had a few notable moments in your Earth caving experience. You were lucky enough to help map an additional few hundred meters of the famous Son Doong cave system in Vietnam. And spending a few weeks on a team mapping some of the underwater caves in the Yucatan was very formative. But perhaps your fondest Earth caving memory is finding a previously unknown cave entrance in the wilderness of northern British Columbia and being one of the first to map it. While that job is certainly not done, you were able to help find five kilometers of unexplored tunnels, chambers, and caverns. It was harrowing, a little scary, and absolutely addictive. Where else can you push the boundaries of caving? Where else can you stand in a place that, perhaps, no one else has ever stood? The answer: a cave system in space!

In your time, space travel is so unremarkable that it was easy to join otherworldly expeditions. Both the Moon and Mars, while quiet and cooled today, were once home to global volcanism like Earth. The remnants of this past activity are obvious both on and below the surface. Ancient lava tubes that once held underground rivers of magma wind their way through solid rock. These lava tubes are excellent places to build human habitats now, as they help protect the inhabitants from solar and cosmic radiation. Stepping foot into one of these lava tubes truly scratched that itch you so long had tried to reach. You spent three whole months on Mars doing this, a very surreal place to live, both Earth-like but somehow not. You even took a moment to lay back and watch the clouds roll by.

That was the crowning achievement of your career, up until you were offered a chance to be part of the second Hyperion caving expedition. It's taken a year to get out here, but now that you're staring through the window down at Hyperion's iconic surface, it feels like the blink of an eye. From this vantage, only about 1,000 km up, the surface looks almost sponge-like, or perhaps like a piece of coral. This is because the surface is covered with thousands if not millions of overlapping craters. These craters have steep walls and make it look like it's completely riddled with holes.

With your Moon or Mars expeditions, there were many teams who had visited these sites before you, but with Hyperion, there has been only one previous team. Thus there is much exploring yet to accomplish. So, this is big. And for years the caving circles have collectively hoped that spelunking on Hyperion would be unlike anything in the solar system; you can barely contain your excitement and hope that this is true, as your spacecraft touches down.

Why does the community think it could be amazing caving? This is because of Hyperion's measured porosity. If you measure Hyperion's size, about 270 km across, and its mass, you end up realizing there is not enough matter to fill that volume. Another way of saying this is that Hyperion is about 40 percent porous, or 40 percent empty space. This could mean a moon-wide cave system to explore, riddled with tunnels and caverns, and interconnected in infinite ways. But while we know it's 40 percent porous, we don't know how big those empty spaces are. That 40 percent porosity could just be trillions and trillions of small and unconnected little holes no bigger than the size of your fist or even a tiny bubble in a fizzy drink. Yes, the radar measurements taken by orbiting spacecraft indicate the first kilometer down probably does have some kind of open spaces that are likely large enough to fit humans, but there's a difference between radar measurements on a computer and walking through it yourself.

After landing, getting acquainted with the hab, and unpacking, it's impossible not to want to get started. Without an atmosphere, the environment reminds you of your time on the Moon. But you need to be careful; you can't just stroll out onto the surface and dive in. The reason Hyperion is so different from caving expeditions in the past is the gravity, which is hundreds of times weaker than on Earth, or even the Moon. With a gravity field like this, even walking can be difficult, because with each push you leave the surface for quite a while and can tumble out of control. The easiest way to get around is to make use of guide ropes and carabiners, holding your hand and pushing off, sliding along the line until your feet reach the ground again.

Packing up your space suit, multiple canisters of extra air, ropes, clips, climbing gear, some handheld science equipment, and a low-thrust jetpack (just in case), you're ready to go. Step-floating out of the hab, you come to rest on the regolith; the cave entrance is only about 100 meters away. You pull yourself down the guide ropes, your gear on a rope clunking along behind you, periodically clipping-unclipping-reclipping your biners around anchored stanchions. Saturn looms on the horizon and the ground is sharply lit. Your anticipation grows.

Finally, the rope leads to a small opening—it couldn't be larger than a meter or two in size—carved into the side of what must be a crater wall with a 30-degree slope going hundreds of meters uphill and downhill. This is it! Despite the incredibly different environment, the feelings are surprisingly comparable to when you were a kid on hands and knees crawling through holes: excitement, nervousness, anticipation.

You take one last look up at Saturn, click on your head lamp, and pull yourself through the opening. The tunnel remains about the same size for a few minutes of traversing but quickly gives way to a large chamber, probably the size of a house. Now called chamber one, this is where the first team has light rigs set up, and some equipment stored. Their trip to Hyperion ended here, just ten meters or so into the surface. But this is good news! Perhaps there are more chambers this size? Maybe Hyperion will live up to the hype.

Looking around, you see many cracks and crevasses. The room is irregularly shaped, with a very jagged ceiling and walls. You make note of a few possible tunnels too small for you, maybe large enough for your arm, a couple on the ceiling, a large crack leading from ceiling to floor. Walking up to this crack you shine your light along its length but can see only a short distance. Chipping away at the wall with a hammer, loose material falls away. Hyperion is primarily made of water ice and looks like glacier ice back home. Inside the "rock" you find many hundreds of small cavities, almost like pumice: porous and not in the way you hoped for.

On the far side of chamber one, opposite the entrance, you find another opening leading deeper into Hyperion. It's been marked by the first team as the way forward, and you agree: there's nowhere else to go.

This tunnel is smaller than the entrance, but still comfortable, a few feet on each side of you. Plenty of space! Reaching out to jagged ice chunks, you pull yourself forward, twisting and turning. All the surfaces are jagged and coarse, and could easily catch on your suit, ropes, or equipment. While there's plenty of room, the clunky space suit doesn't make things easy. But, on the upside, the low gravity makes this feel almost familiar, like you're back in the Yucatan scuba diving through its vast underwater cave network.

Occasionally, you place an anchor and connect your guide rope, so you have a trail back out. About 50 meters into the network you reach a fork in the road. Excitement! Glancing to the left, your headlamp illuminates a very thin passageway heading off into the unknown. It looks too thin for all the equipment you have dragging behind you, but you make a mental note to come back. You can probably make it if you remove your air tanks. To the right you can fit with what you have now, but the walls are closing in the farther you move forward. This tunnel is getting smaller, which tempers your excitement. Luckily, you spill out into a much larger area, probably the size of a bedroom. Success! You anchor your guide rope and check your distance, now 100 meters into Hyperion.

You dub this chamber two, and it's about the size of a bedroom. As you inspect this new room, you find there are a few passageways branching from here. A smile grows across your face: Where to start?

Over the next week you start methodically and precisely measuring and mapping the tunnels you find. You return to the fork in the road from day one, but it went only about 20 meters before squeezing to a pinch. From what you can tell, chamber two is where the network branches off, with many tunnels heading in many directions. You follow them all with excitement, only to find they slowly narrow and then eventually cut off, forcing you to backtrack. Back in chamber two again, you sit at the opening of the last passageway left. If this one goes nowhere, you'll have to find a new cave entrance elsewhere.

As you've done ten or fifteen times now, you set your anchors and begin down the new path. This opening is especially narrow, perhaps the reason you left it to the end. Roughly circular, it's just wide enough to fit your suit and drag your equipment behind you. You reach forward and, hand over hand, pull yourself through and down farther into Hyperion. Your face mask almost scrapes the icy passageway. Occasionally, the tunnel widens a bit. It's at one of these spots you decide to drop your pack and equipment; it's too narrow for anything but the essentials at this point.

Moving further forward, you check your distance and realize that after a few hours of this tight tunnel, you're almost two kilometers in! This is the deepest you've yet traveled into Hyperion. Looking ahead, your headlamp casts odd shadows in many directions, but you notice something in the wall. You push yourself toward it and find a small hole, maybe the size of your face, a window into another tunnel or opening just on the other side of the wall. You try various angles and adjustments to see as much of the other side as possible, and you can tell that the cavity on the other side is probably only one or two meters across, just barely big enough for a human. But it's completely sealed off other than this window. You make a note on your computer and move on. You wonder how many more of those inaccessible cavities exist throughout the moon?

As you move further forward, you start to find many branching tunnels and many more windows off into strange and hidden places. You are cautiously optimistic: it's possible that the hopes and dreams of the spelunking community back home have come true. This could be a caving paradise! It begins to feel as if you have a limitless number of tunnels, passageways, and rooms to explore.

Mapping and charting as you go along, you find yourself marking all the branches and rooms and chambers that you hope to come back to over and over during your time here. As the days go by you find you've mapped nearly 15 km into Hyperion and are now at a count of eleven chambers! This couldn't be more of a success.

You're starting to near the end of the guide ropes you brought for mapping your path, so you know it's almost time to turn back. Pulling yourself along, you realize this is the tightest squeeze yet. Stopping for a moment, you unclip your jetpack and unhook your oxygen tank to drag behind you, making yourself narrower. You grip ahead with your hands, through a tunnel barely bigger than your space suit. Just a bit farther.

It's hard to get a good look ahead because it's so narrow, but the view forward is a little confusing: it just looks black. Perhaps a new chamber? Normally, you'd see the other side of the chamber from here. A few more meters and your hand reaches the point where the blackness begins, and you reach out into empty space. Huh. This is different. Pulling your head through the hole, you see there is floor around you, so you clamber out onto it and realize you are in a chamber, but your headlamp is not strong enough to illuminate the entire way across. So you pull your high-powered light that was dragging behind you and shine it out into the void. You can see more of the ground around you, yes, but still can't see a ceiling or a farther side. Your eyes widen, this must be big.

Pulling out your laser range finder, you point it out into the abyss of blackness. Let's see what we have here. The readout blinks a few minutes, then displays 3,128 m! Pointing in a different direction, it reads 15,322 m! Incredible. You've dragged yourself through tunnels and chambers of ice and stepped out into a cavern large enough to house a small city.

A quick mental calculation leads you to the conclusion that, for a moon that is roughly 250 km in diameter, and has a porosity of 40 percent, there can be only a few chambers of this magnitude in the entire moon!

Floating down to a sitting position, you take a moment to sit in the emptiness. The ice here is pristine. There's a part of you that wants to etch your name on an ice boulder to leave something of yourself here and to celebrate this first arrival. One whose name was writ on water indeed! But somehow it just doesn't feel right. A subtle beep pulls you out of your reverie, alerting you that your oxygen level suggests that you get moving. There's a lot of cave to double back on, and just enough oxygen to do it.

★★

Hyperion is one of the many moons of Saturn.[1] It orbits at an average distance of 1,500,930 km from Saturn, which is just a slight bit farther than Titan, which orbits at about 1,200,000 km. Hyperion is one of the places that reminds us just how weird things can get. For example, it is the only moon in the solar system that has a chaotic rotation. This means that it doesn't rotate simply around an axis like almost everything else. Instead, it tumbles randomly, always changing its current axis of rotation. Thus it has no preferred spin axis. The reason for this is a combination of a few things. First, its orbital eccentricity is about 0.1, which is rather high. This means that Hyperion has an elongated orbit that continually brings it close to and then takes it out far from Saturn. Second, Hyperion is oddly shaped, which means that as it gets closer and farther from Saturn, Saturn's gravity is always pulling on its surface unevenly. Finally, Hyperion is also in an orbital resonance with Titan. In fact, there is a continuous tidal tug-of-war between Saturn and Titan. The combination of the orbit, shape, and gravitational influences from both Saturn and Titan has led to a situation where Hyperion is slowly tumbling chaotically.[2]

## A SPONGE IN SPACE

Hyperion just looks different from most places. It truly does look like a sponge or a piece of coral, or maybe a piece of pumice. It almost feels as if it would float on water.

Slightly related to its visual appearance, the truly mind-bending feature of Hyperion is its measured porosity. In geology, the term *porosity* refers to how much empty space there is in a rock or object. Hyperion's porosity is greater than 40 percent, which means that over 40 percent of the moon is empty space. And this is why we chose Hyperion as a fun place to investigate. With that much empty space inside a solar system object, what would it be like to dig down and take a look?

To calculate the porosity of an object in space, you need two numbers: the object's total volume and how much of that volume is void of rock. Then you divide the latter by the former. You end up with a percentage that tells you how much of the object is, well . . . not rock. Easy, right?

Easy on paper, but tough in practice. How do you measure the volume of an object, and just how do you measure how much of that volume is not rock? The answer to both questions is: measure its density. Density is a calculation of the mass of an object divided by its total volume. Let's start with how we get volume.

The equation for the volume of a sphere is $V = \frac{4}{3}\pi r^3$, where $r$ is the radius of the sphere. Unfortunately, Hyperion is decidedly not a spherical object. In fact, it's one of the largest nonspherical moons in the solar system.[3] To figure this out, back in 2005, the Cassini spacecraft made a few close approaches to Hyperion, flying to within 600 km of the surface to make a series of measurements of both its surface and shape.

Cassini bounced radio light off the surface and measured the return (not unlike how the surfaces of Venus and Titan were mapped). Using that data, scientists can carefully construct a three-dimensional model of the object.[4] With this information, Cassini was able to measure Hyperion's shape to be something akin to a potato, with three axes measuring 410 km × 260 km × 220 km. Using this model, the total volume can be estimated.

With a measurement of the volume, the next step is to measure the mass. But how do you measure the mass of an object floating in space? Mass is a measure of how much matter makes up an object, and matter creates the force of gravity. If there is a lot of matter in the moon Hyperion, then it will create a strong gravity field, and if there is not a lot of matter, then its gravitational pull on other objects will be less. So, to measure the gravity field of Hyperion, and therefore its mass, scientists use the trajectory of Cassini. By flying Cassini by Hyperion, and then watching to see how Cassini is pulled by Hyperion, we can measure the amount of mass. This number comes in at approximately $5.6 \times 10^{18}$ kg, which is about 10,000 times less than our Moon (or about 10,000 times the mass of Mount Everest).

Okay, now with measurements of both volume and mass, we can calculate density, which is mass divided by volume. Hyperion's density is about 544 grams in every cubic centimeter of material. And here's where it gets interesting: that calculated density is about half that of water.

If we assume Hyperion is mostly made of water ice, and some trace amounts of other ices, then we've run into a problem. If the material is clearly mostly water, but is measuring a density of about half that of water, that means there would have to be gaps inside to allow it to fill the entire volume. In other words, Hyperion would have to be porous. This means that it's not just layers upon layers of ice; it must have a bunch of gaps throughout its structure, like a sponge or like Swiss cheese. As described

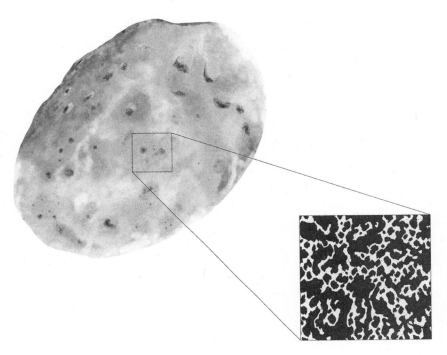

Figure 13.1
Hyperion's density is so low that it may actually be a rubble pile of blocks held
together by gravity. Naturally, these blocks won't always fit together perfectly,
leaving gaps in between that our protagonist explores as caves.

above, the number of gaps in the structure is called porosity, and Hyperion
appears to have about 40 percent of its volume taken up by empty space. In
figure 13.1, we imagine what this could possibly look like.

We have already encountered an object that has a similar interior structure: Bennu. In chapter 10 we looked at the asteroid Bennu, and how both
observations from afar and samples of the surface show that it's more like
a pile of rubble than it is a solid rock. Hyperion is similar to this, only it is
made of ice, and it is likely a little more rigid than Bennu since ice can be
readily sintered together. Note the size difference. Bennu is less than one
kilometer in size, while Hyperion is hundreds of kilometers. The gravitational field at Hyperion is holding these chunks and boulders of ice together
much more firmly than the rocks of Bennu.

While we do know much of Hyperion is empty space, what we don't
know is how big the empty spaces are. That unknown can lead to some

fun Fermi problem–like estimates. Given the measured axis sizes, we can estimate the total volume of Hyperion to be $V = \frac{4}{3}\pi(410 \times 260 \times 220) =$ 9,953,337 km³, or let's round to about 10 million km³ (that's about 100 times less than Earth's ocean's volume). Of course, 40 percent of that volume is empty space, so that means about 4 million km³ of empty space in Hyperion.

If we assume the gaps are roughly the size of a house (10 m × 10 m × 10 m), then each gap is about 0.000001 km³. How many gaps the size of a house can we fit inside 4 million km³? Trillions!

What if the gaps were the sizes of cities? Let's say 10 km × 10 km × 10 km; that's a gap size of 1,000 km³. How many of those can we fit inside 4 million km³? Thousands!

What's most likely is a range of different sizes from small to large, which leads us to imagine an interior structure of Hyperion littered with small holes, tubes, tunnels, small rooms, and caverns. Some connected to each other, and some sealed off. An endless series of caves to be explored. A spelunker's paradise.

We can't leave the Saturnian system without taking one last look at Titan. This will be our last flight of fancy around Saturn before we cap off our journey at the edge of the map.

## CAFFEINATED FLIGHT ON TITAN

It's morning at the Huygens Historic Landing Site Visitor Center on Titan as you pull your unpressurized vehicle into the parking lot. You wouldn't know it from the sky, which is the usual dark yellowy-orange color common to Titan's day, which is nearly as long as 16 Earth days. You give a glance around before heading in—it looks like you're the first to arrive, well before any of the tourists. That is, if any show up on this day, well into the slow season.

After cycling through the airlock to the staff area of the museum, you remove your mask and goggles. They stay clipped to your jacket rather than placed in your locker because you're planning just a brief pit stop here before you head out on your rounds.

But first, coffee.

You drop into the kitchen, take off your bulky outer coat, and start the ritual. You release fresh beans from a tiny vacuum-sealed pouch from Earth. Bringing the opening close to your nose, you close your eyes and savor the aroma for a moment. You recall reading that on the old-fashioned airliners of Earth, extra salt and sugar were added to in-flight meals to make them palatable to passengers, whose sense of taste and smell was dulled by the effects of low pressure. That's not a problem here on Titan where the atmospheric pressure is 50 percent greater than it is on the Earth. Some recent immigrants from the inner solar system even claim that everything tastes and smells more vivid here on Titan. Of course, when you've been dining aboard a cramped spaceliner for months, you might imagine any planet-side food would taste exceptional.

As someone born here, who has always known only this dim hazy sky and these soft hills, you wouldn't personally know the difference. But it's nice to think that you're getting a special experience, a small but private pleasure.

Real coffee is a luxury out here on Titan, and everyone at the museum pitches in to buy the coffee for the daily brew, a task entrusted to you. Two factors make you perfect for this critical task. Not only do you like to get in early, but you have a quiet, meticulous nature that allows you to focus on the task. Coffee brewing may be an art, but on Titan it's equal parts science.

Grinding up the beans is relatively straightforward—any burr mill will do the trick. Fresh water is also very easy to obtain—the surface of Titan is built of the stuff, so buildings here essentially just mine their water from the ground nearby and filter it, or get ice delivered. No water mains are necessary (though wastewater disposal is a whole other problem). Next, that water needs to be heated precisely to 93°C, the same as it does in the inner solar system. However, water boils on Titan at 112°C on account of the pressure. Using water that hot would scorch the grounds, giving them a burnt flavor. There's also a danger of overextraction with drip and pour-over methods, since the low gravity here on Titan means that the water flows much too slowly through the grounds unless the individual coffee particles are so large that much of their flavor is wasted. There's no issue with espresso machines, which use pressures far higher than those here on Titan, but it can be a challenge to make enough for a group this way. Finally, for such precious cargo, boiled "cowboy coffee" was dismissed out of hand.

With a little trial and error, you have found that you can get the best results using a vacuum brewer with some slight modifications. Pure water in the bottom flask is heated by induction. As the water gets closer to its boiling point, the increasing water vapor pressure in the flask pushes the remaining liquid up the glass siphon tube in the center, through a paper filter and into the upper chamber. That water is near to boiling and far too hot, so the next step is to add some fine, cold quartz sand.

The sand accomplishes two tasks. Like a whiskey stone, the sand cools the water without diluting it. The more sand is added, the more the mixture is cooled, creating a vacuum in the lower flask as the steam condenses back to liquid. Second, the sand slows the return of the now cooled water from immediately descending back into the siphon in response to that vacuum. The finer the sand, the longer it takes for the water to be sucked back into the lower flask.

From this point onward, it's all in the timing: reddish-brown coffee grounds go into the prepared bath and brew in water that is being slowly drawn downward under pressure into the lower chamber. You slowly stir the grounds to speed the extraction and to prevent them from accumulating on top of the sand. A swan-necked kettle stands ready to adjust the amount of water or the temperature upward. This could be a nerve-wracking task, but you are well practiced and carry it out with calm detachment, adjusting the process in a natural way based on nearly subliminal cues.

As you brew, you look up and afford yourself a glance out the windows at the Huygens probe, sitting on the pebbles in the inner open courtyard of the museum, right where it landed, so many years ago. Somehow, it's a sight of which you never get tired.

Finally, you have several liters of fine coffee in the lower flask. Disconnecting the upper chamber and discarding the expended puck of sand and grounds wrapped in the paper filter, you draw off a cup for yourself. The coffee is delicious. No need for cream or sugar. You couldn't write down exactly how you accomplished this perfection even if you tried.

Fueled up, it's time to get to work. On your way out of the kitchen, you place the flask into a pre-warmed insulated holder to await the arrival of your colleagues. You'll say hello to them when you get back in a few days, but you prefer to have your mornings to yourself, especially when you are about to head out to the park.

Grabbing your coat, you go down to the locker room and trade out the heavy insulator for a thinner garment with heating elements spiderwebbed throughout and a helmet with a clear faceplate and a heads-up display. Satisfied with the suit's readouts on the flow of oxygen and its battery levels, you head for another airlock that will bring you to the museum's equipment shed.

The museum itself is shaped like a donut on stilts, with the relic of the Huygens probe in the middle. The reason for the stilts lies with the temperature difference between the ground and the building itself. The ground hovers around the surface temperature of the planet, a chill 94 K, or −179°C, while the interior of the museum remains at a comfortable 20°C. Though well insulated and as thermally isolated from its foundations as possible, the building will always have some waste heat that will need to escape. Couple that with ground that can melt and you get an outer solar system version of Arctic architecture. Indeed, like many high-latitude pipelines, the museum has radiators to release as much of the waste heat as possible into the air without transmitting it into the ground.

To protect the landing site and the spacecraft upon it, a transparent roof has been mounted overhead, covering over the middle of the donut. Meanwhile, a berm of ice blocks surrounds the visitor center to prevent sudden deluges of methane rain from washing the probe away. This had been a problem before the construction of the visitor center: previous floods had relocated Huygens from where it landed. Unlike similar visitor centers on the Moon and Mars erected to memorialize other spacecraft relics, there were debates on Titan as to whether the visitor center ought to be located at Huygens' final location or whether the delicate spacecraft should have been brought back to its original landing point.

However, all of this scarcely matters to you. Your interest and objective is the environment far from this particular building. This place is simply your jumping-off point.

The equipment shed is located on the external curve of the building, opposite the parking lot. It's an unpressurized space where equipment for the park rangers, like yourself, is kept. It's the sort of space that might be described as musty if anyone could smell it long enough without gagging on Titan's toxic soup of an atmosphere.

This morning, you log out a pair of articulated wings and a small hand-held propellor rig in a protective cowling to provide vectored thrust, not unlike a diver propulsion device or thrust pod. Then you climb the ladder to the visitor center's roof.

Once up on this purpose-built platform, you get ready for takeoff. First, you clip the propeller rig to the front of your suit. Next, you attach your wings. The wings come with a harness that is worn over the shoulders and through the legs, like a longer, more intricate version of a climber's harness. You make certain that you are strapped in tight. The wings can lock into place, lying flat against the back and extending out at the sides to a point about a foot beyond your hands at arm's length. In this configuration, your arms are free to move about and to do other things, for instance, use an instrument or camera or employ the propeller unit.

However, with a command, the wings can lock into attachments on the arms and the articulations release to allow a human-powered flapping motion.

It's a very specialized Titanian device, born of the innate desire for human-powered flight and the joy of achieving that long-held dream. Many artisans have spent countless hours on the craft of these wings, perfecting the materials used, the shapes of each element, and the articulations between those elements. The result is that the whole flying rig is lighter and more fluid and intuitive to use than one would imagine possible.

Indeed, you've been flying ever since you can remember, and in some ways it's a more natural way to move about than the shuffling steps required on a world with similar gravity to the Moon's and much thicker air than at sea level on Earth. You can't imagine visiting one of those higher-gravity places; indeed, if you were to travel to the Earth or Mars, having grown up on Titan, your skeleton simply would not support you. Just to walk around you would need your own powered suit to help you do the work needed to oppose what for you would be a crushingly high gravity.

But here, you feel light and free to explore wherever you want to go. And today, you have a job to do. Best to get at it.

You set the wings to articulate and take a practiced jump from the platform, thrusting yourself forward and upward. Gravity tries to pull you down, but a few flaps of your wings thwart its best efforts. At first you head out beyond the museum with rapid and powerful strokes to give yourself room to maneuver, then you curve your trajectory back around toward the museum. Those radiators that keep the museum's foundations solid also provide a gentle updraft that you can use to gain altitude in an upward spiral.

Not everyone uses this technique. Some rangers go straight for the propulsion pod, to lift them skyward. But you've always enjoyed the flapping start—it's just you and the wind working together, unmediated by any other person or technology.

Tourists from the inner solar system often ask, and rightly so, how can wings not much larger than those of a golden eagle hold up a human being? That a person could have the power to maintain flight is something well beyond their experience.

There are two big advantages that a Titanian has over an Earthling.

The first is the atmospheric density. Because of the cold and the elevated pressure, even though Titan's air is made mainly of nitrogen, the same stuff as the Earth's atmosphere, the Titanian air is 4.5 times more dense. That means each stroke of the wings generates 4.5 times as much lift. That benefit doesn't come for free—that wing stroke also generates 4.5 times as much drag, which means that going fast in the air on Titan takes effort. The tourists often report feeling this same trade-off to a much greater extent whenever they swim in water.

The second advantage is gravity. Because Titan's gravity is 7.3 times less than the gravity on the Earth, the same aerodynamic lifting force can support 7.3 times as much mass as on the Earth.

The net result of both effects is that wings that support a five-kilogram golden eagle on the Earth can hold up something 33 times heavier on Titan, say, 165 kg. By that calculation, human-powered flight is not just for elite athletes but for everyone. Indeed, you know that the primary reason people make the trek all the way out to Titan in the first place is to experience self-powered flight—something available nowhere else in the solar system.

Of course, someone who has never flown before doesn't always know how to do it right, and plenty have gotten themselves into trouble. That's where you come in. For the next week, you're out patrolling the Shangri-la planetary park from the Huygens Visitor Center, just south of the equator on the plains all the way up to the Dragonfly Visitor Center at Selk Crater, just north of the equator. Each night, you'll be able to recharge your propeller and have a warm meal and a comfortable bed at ranger stations situated around the park.

It's a journey of nearly 860 km, as the crow flies, and that's too far to flap on your own, which is why you have the propeller with you. You think back to the first flying vehicles in this world. For the Dragonfly spacecraft, at its top speed of about 10 m/s (36 km/h), such a journey would have taken about 24 hours of constant flying. Though you can fly faster and much more efficiently with your fixed wings, typically you maintain a similar pace—it gives you time to think.

Eventually, your spiral takes you up to a pleasant cruising altitude of about 10 km. On Earth, you would be in the stratosphere—the domain of passenger aircraft—with less than a third as much pressure, and temperatures that would have fallen by 80°C or so. But here on Titan, the rate at which the atmosphere gets thinner and colder is much slower. By the time you reach this altitude you have only lost about a third of the pressure and just 10°C in temperature. Up here, the surface below you is still relatively sharp, if a little bit hazy.

But you do have the advantage that you are higher than any tourist's rig will allow them. You'd never see them from this far away, but luckily each has a transponder included and you can overlay their positions on the display inside your helmet. Today, no one seems to be out and about. So, you take a pause, circling, and look around.

In some ways it's a strange scene, suspended as if you are beneath a dim, diffuse sky above a surface lacking much in topography or detail that is recognizable to the human eye. From this altitude the horizon is over 200 km away, but it too is hidden by haze. Here in Shangri-la park the landscape is dominated by dark dune fields in every direction, though you could navigate by the dry arroyos that dissect the landscape if you had to. As a result, there is no sense of scale to the untrained human eye. You could be forgiven for thinking that you were suspended in a small room instead of being a speck lost within thousands of cubic kilometers of open space walled and ceilinged with layers of soft, sheer yellow fabric.

Indeed, it is for this reason that tourists are restricted to flying lower than a kilometer or so from the surface. Higher than that, without the appropriate sensory inputs, it's easy to become disoriented or to lose track of time, like being immersed in a sensory deprivation tank. However, for the experienced flier, like yourself, the feeling you get is one of calm.

Since there are no tourists to check on, you decide to make a random stopover in the wild before heading to the ranger station. The place doesn't have a name and it isn't very specific—just out in the middle of nowhere in a sea of rolling sand dunes, a few tens of kilometers north of the Huygens visitor center.

On your way, you take the liberty of practicing a few aerial tricks—there will be a competition coming up in a few months and you might as well use the time productively. The speeds are fast on a human scale, but slow compared with flying machines elsewhere. You practice barrel rolls, loops, and flat spins. You auto-gyrate downward in a tight spiral like a maple tree samara. Given the power of the propeller, there are even a few Titan-specific maneuvers that are spectacular to behold.

With each trick you give up a little bit of altitude until, at last, you approach the surface once more. With the propeller pointed skyward, you alight softly on the ground and break through the crème brûlée–like crust of the surface in this wild area, far from any path or route. It's likely that you're the first person ever to have arrived at this particular unidentifiable spot.

The only sight for miles around is the linear rows of dunes made up of dark granules of the same tholins that compose the atmospheric haze. You reach down and rub these particles between your fingers. Each one is a fraction of a millimeter in diameter, like fine sand. Here in the trough between the dune crests, when you reach down and break through the thin crust, the sand is saturated with liquid methane that fizzes gently as it comes into contact with your suit and boils away.

It reminds you exactly of the coffee grounds you brewed earlier that morning. A whole monochromatic landscape made up of coffee grounds in various hues of reddish-yellowish-brownish-black.

As you climb up to drier ground, you consider that these particles began life at the top of the atmosphere in the light of the Sun where ultraviolet light broke apart molecules of methane and nitrogen gas. The products of this photolysis were solid, not gaseous, but were irrevocably chemically changed from what came before such that they could never evaporate away like a cloud does. The particles were so small that they fell very slowly in Titan's thick atmosphere, bumping into others like themselves, in some cases bouncing off, but more commonly aggregating together into immensely complicated fractal patterns.

Up near the Sun, the winds are quick, circling the planet far quicker than it spins on its axis in a process called super-rotation. But as the growing particles became larger, they fell faster, eventually falling deep enough in the atmosphere that the Sun was nothing more than a memory, now evidenced only by a slightly brighter yellow in part of the sky. In the lower part of the atmosphere, things were calm and quiet. Here, the temperature changes by no more than a degree over the year and the winds are constant in direction and intensity. When the particles approached the surface, those winds swept the tholin particles into organized ridges, forming the largest sand sea in the solar system with whorls outlining every mound and crater.

"Sea" is the right name for this landscape. The standing wave–like frozen formations you can see from the top of this nameless dune evoke a watery swell for the tourists. But the truth goes deeper. The methane molecules from which the tholins formed came from a methane ocean that once covered all of Titan. Over time, that methane evaporated and traveled upward to be converted into tholin at the top of the atmosphere. These coffee ground–like sand grains are all that remain to tell the story of that vast and ancient sea, frozen into an imitation of their former existence.

And isn't that true of any great journey? That we return to where we started, only to find that it is ourselves who have changed?

There is pride in this land for you, and you count yourself fortunate that you can spend your days exploring and experiencing it in ways large and small. Soon enough, your footprints up to the crest of the dune will wear away—they don't fit the pattern being sculpted by the wind and will be replaced by new tholin grains. No evidence will remain that you were ever here. But you can enjoy the view and contemplation for their own sake.

You take one last look around, then, switching on the propeller, you alight into the sky.

★★

It may have seemed weird to be sailing on one of Titan's hydrocarbon seas of methane and ethane back in chapter 7, but now that seems almost normal in comparison to flapping through its atmosphere like an oversized bird. Though difficult to picture, this is entirely possible due to the unique features of Titan. Luckily, this story is paired with something more commonplace to keep you grounded.

### THE EXOTIC WRAPPED IN THE FAMILIAR

What could be more common than the daily task of making coffee? It's a task shared by well over a billion people every day. Yet even a habit so simple and taken for granted changes when you transport it somewhere else in the solar system. As we all learned at an early age, water boils at 100°C. In fact, that's the very definition of the Celsius scale.[1] When the $H_2O$ molecules in the liquid water reach such a high temperature, they can break their bonds with their neighboring molecules, and fly away from the liquid in which they are submerged. This manifests itself as a whole bunch of bubbles in a rolling boil. However, there is a slight caveat to this well-known fact: water boils at 100°C only if you're boiling it in a place where the surrounding atmospheric pressure is the same as Earth's atmospheric pressure at sea level. That's about 100,000 Pascals, or we also use 1 atmosphere as a unit of measure to mean "the pressure at Earth's sea level."

Remember, as we discussed in chapter 5, what we experience as atmospheric pressure is a measure of the collective strength of the countless collisions of the atmosphere's molecules. This means that if the pressure is higher, there are more atmosphere molecules pushing down on the liquid water you are trying to boil, making it harder for them to fly away, and preventing bubbles from forming. Thus, in order to boil the water, a higher temperature is needed to overcome the pressure.

Figure 14.1
Even something as simple as boiling a kettle happens at different temperatures in
different places. What does the higher boiling point of water that we have shown for
Titan tell you about the atmospheric pressure at the altitude we considered?

As we discussed back in chapter 7, not only does Titan have an atmosphere, but it also has a thicker atmosphere than Earth's. Titan's atmosphere
is 1.5 atm, or about 50 percent higher than Earth's atmospheric pressure at
sea level. Thus, you would need a higher temperature to make water boil
on Titan.[2]

You can try the same experiment in reverse if you visit a high mountain
top on Earth. There, water will boil at much lower temperatures than at sea
level. This would be a more severe challenge than the one our Titan ranger
faces, for explorers who are aiming to extract their favorite brew from coffee grounds. In figure 14.1, we take this one step further: water boils at
100°C at Earth's surface, but surprisingly requires 112°C at Titan's surface.

## HUMAN FLIGHT

On Earth, even with technology, flight for human beings is difficult. Human-powered aircraft and jet packs have been developed that make flight marginally possible, but neither technology is truly convenient in the way that it would be on a moon like Titan. On Titan there are two factors that make flight that much easier: the density of the atmosphere and the reduced gravity.

Recall from chapter 1 how to calculate surface gravity. Titan is both a small object (5,150 km wide) and not very massive, which leads to a surface gravity of about 0.14 $g$, only slightly less than our Moon's. It's easy to believe that this would make flight much easier. There's less gravity to fight against.

But just as important is the increased density of Titan's atmosphere. Since it is 1.5 times thicker than Earth, there are that many more atmospheric molecules against which to push when flapping your way into the air.

Taken together, each turn of a propeller, flap of a wing, or spurt from a jet engine has a much stronger effect on your lift, as described in the story. It may be that Titan is the only place in our solar system where humans could truly achieve flight as we've always imagined.

Additionally, the same advantages make the risks of a problem and a fall somewhat less severe. In chapter 11, we learned that terminal velocity is a consequence of the drag created by the atmosphere you're falling within. On Earth, terminal velocity for a skydiver approaches a lethal 200 km/h. We also saw in chapter 11 how terminal velocity approaches a truly nightmare-inducing 2,000 km/h on Jupiter. But on Titan, since the gravity is lower, and the atmospheric drag is much higher, you would hit the ground at barely more than 30 km/h. Painful? Yes, but not deadly. As a result, flying might be a preferred mode of transportation for explorers or even commuters on Titan. In figure 14.2 we compare some of these terminal velocities.

An additional advantage of Titan is that the densities seen near the surface persist to rather high altitudes. Regardless of which atmosphere in the solar system you're looking at, the highest pressure is always at the bottom of the atmosphere (i.e., the lowest point), and the pressure "decays," or decreases, the higher up you go. This is why Mount Everest has such a low pressure at its summit. This rate of decay of pressure happens in all atmospheres and can be described well by the *pressure scale height*.[3] That scale

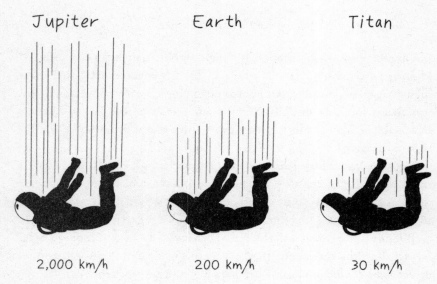

Figure 14.2
As a human being (or any other object) falls in an atmosphere, they will pick up speed until the force of their weight is balanced by air resistance. This happens at drastically different speeds in different atmospheres.

height is the altitude you need to reach for the pressure to fall to a factor of about one-third. On Earth, it takes about 8,500 meters for that to happen, just a little less than the altitude gained by a trip up to the summit of Mount Everest from the Indian Ocean. On Titan, you'll need to rise all the way up to 20,000 meters before that amount of pressure decrease occurs. As a result, Titan has much more navigable altitude to explore.

## ANCIENT TITAN

All worlds, even places that seem frozen in time, like the Moon, go through events that utterly change their appearance. Deep geological time spans are what allow rivers to carve canyons out of solid rock, after all. It is difficult for us humans to come to terms with 100 million years as more than a number on a page, but using the power of geology we can certainly infer how a world has changed from the start to the end of such a span of time.

Due to a variety of geological, biological, and atmospheric processes on the Earth, our planet's surface is changing so quickly that many of the features we observe are very recent, just hundreds, thousands, or perhaps a few million years old. Except for a few cases, you really need to go looking to identify any impact craters on our planet at all, as most have been eroded away, buried, or otherwise destroyed. By contrast, on Mars you can barely kick a rock and not hit something that has sat out under the stars for a billion years or more.

Titan is in the middle of that range. Craters are filled in by the shifting tholin sands that are created when methane in the atmosphere is broken down by ultraviolet rays high in the atmosphere. This means either that the methane we see today is being replenished or that there was a lot more on the surface in the past: perhaps much larger global seas of methane a few hundred million years ago, as the daydream discusses. Now, all that remains are the lake districts near the poles and vast equatorial sand seas made up of the processed remains of those hydrocarbons, the echo of those ancient waves eerily preserved in the dunes.[4]

# PLUTO, THE EDGE OF THE MAP

There is some magic that keeps drawing you back to Pluto. It seems silly—this world isn't even a planet anymore, at least not officially. But isn't there some romance to the idea that, for 76 years, this was the edge of the map of the solar system?

You're not nearly so old that your childhood self needed to learn the names of those canonical nine planets. You never needed to make foam ball models of an overly simplified version of the solar system. Indeed, you wonder what those grade school students did with their mnemonics after the "demotion" of Pluto? Did Mary still Value Easy Money and Judy still Save Up Nine Pennies? Or had that group just given up after one too many attempts to modify the system? With luck, by the time they arrived at "My Very Educated Mother Cannot Just Serve Us Nine Pizzas, Hundreds Must Eat," some had realized that names are just an approximation of reality. Rote memorization can get you only so far in understanding nature.

But, in the same breath, the mystics who thought that you gained power over something by learning its true name might not have been entirely wrong. Instead, once named, that thing gained a sort of power over the mind of the person in which the name resides. Haven't you ever been kept awake in the darkness of the night, turning over a concept in your head? It's like a rock in your shoe: something that draws your attention, something that won't be ignored no matter your efforts to divert your focus elsewhere.

Pluto is like that for you, and why not? Here is a planet (yes, you still use that word!) made of materials that would simply evaporate into thin air if brought as close to the Sun as the Earth. A planet that seems inert most of the time but comes to life for a few decades every 248 years when its eccentric orbit brings it close enough to the Sun. A place with methane snow–covered peaks, icy dunes, a layer cake atmosphere, nitrogen plains slowly evolving with a glacial boiling, and spectacular floating water iceberg mountains!

Have a heart for Pluto!

Still, you know this is a private obsession. Most of humanity has moved on. Most prefer to huddle near the Sun, basking in the campfire of human culture and history. Others are drawn to the riches to be made extracting resources that are plentiful near the asteroids or giant planets. Still others are off on their own missions far into the Kuiper Belt and Oort Cloud. Here, they search for the fleeting acclaim and glory of pushing out that endless frontier, establishing a new record that will persist for a short while.

As a result, so far as you know, you have Pluto as a playground all to yourself. You can still remember when you first arrived as a young adult. You and your colleagues had set down at the edge of an unnamed high-altitude network of ridges, not too dissimilar from the better-studied Tartarus Dorsa on the edge of Pluto's "heart." At this altitude, methane snows out of the atmosphere and the rays of the Sun have sculpted this material into a series of repeating ridges called penitentes that are hundreds of meters tall and a few kilometers apart. It's astounding that this could happen since, viewed from here, the Sun is just the brightest star in a perpetual inky-black sky. That cold disk, about the same width as the rings of Saturn as viewed from the Earth, provides 1,500 times less warmth, on average, as it would back home.

After the others had gone to sleep, you quietly headed out on your own to hike in those hills. You traveled far enough, crossing enough ridges, that you lost sight of your camp and imagined yourself utterly alone with the snow as it glinted under the dome of the sky. Each ridge caught the illumination in a regularly repeating and seemingly endless pattern of light and dark. It reminded you of being a child, sneaking out after dark and climbing to the top of the roof of your suburban home. Laid out before you, in a surprisingly similar way, were rows upon rows of rooflines, each catching the moonlight. The quiet of the sleeping city, muffled by the winter snow, fooled you into thinking that this was a view created just for you.

It had been thrilling, in the child-like way that something you haven't been expressly permitted to do is thrilling (and that is, you must admit, just a little dangerous). It had been a long time since you had felt that kind of peace, that sense of time crystallized and standing still. As a child, you imagined yourself jumping from rooftop to rooftop in that scene, and the same feeling came upon you in the Dorsa. But your adult self had come prepared with a better idea.

You had carried with you a roll of low-temperature plastic with a few convenient handholds that you now flattened out on the ground. By laying down on that sheet, you could push off gently with your feet and start sliding down the slope. As the terrain started to pass by with terrifying speed, you still distinctly remember wondering if you were the first planetary tobogganist. That is, when you had time to wonder anything at all, so powerful was the visceral pleasure of flying down the slope.

Your momentum had even carried you halfway up the next ridge before you rolled off the plastic, giggling in the snow like you had not done in years. Good thing you had turned off your communications system for the ride, or you might have had some explaining to do the next day! You checked your suit's instruments and saw that the whole trip had taken you just under four minutes and your speed had topped out at just under 90 km/h. That would be considered an exciting toboggan run on the Earth and very achievable for even a minor daredevil. It's amazing to think of all the strange parallels to home that exist in this alien place. Not least of which is that the ridges you had been sliding on here on Pluto were also carved by the same physics as the snowpack on Earth; they're just much, much smaller back home.

Emphasizing the larger-than-life scale of this place, your last memory of this night to remember was lying there staring up at Pluto's enormous moon, Charon. Though Charon is nearly three times smaller than the Earth's Moon, it's about twenty times closer. As a result, the crescent in the sky is more than 3.5 degrees across, or seven times the diameter of the Moon under which you grew up. That means the area of the full disk of Charon on the sky could swallow nearly fifty full Earth Moons. No wonder Pluto can seem like a place out of a children's book.

After your night out, you were simply hooked. You began looking for any way to get back to that planet, and luckily you found it. You and your colleagues had been visiting Pluto because the active part of its year was about to begin. The last time humanity visited this place, that active time was well underway. As such, there were some research questions about what a couple hundred years in the deep freeze might mean for the landscape and how things would change over the coming decades as Pluto revived. But the answer to that question was not of much interest beyond academic circles. Even within those circles, many were hunting for the next shiny thing— what was the point in rehashing discoveries already made by New Horizons?

But if the request was simple enough, and the resources needed were small enough—say, just one person stationed on the planet for a few decades who could relay measurements back to the scientific community on Earth—the ask might just match the interest. Following a few psychological tests that proved you would not go crazy spending so much time by yourself, the grant review board agreed.

You've been here ever since, maintaining long-term equipment and performing tests. Some of those you have dreamt up, but plenty of experiments were suggested by others. Though you get back to the inner solar system only a couple of times a year for conferences, you don't feel hard done by. You keep in touch with friends and family at home and your career is going well, even if you aren't the top researcher in your field. Indeed, you feel good about what you're doing with your life because you chose this path, the one that led you right to where you want to be. You've come to know this place like a friend and that means that you're able to pick up on even the smallest changes, a valuable asset to your backers on Earth.

You've traveled widely on Pluto and it feels as if you've been everywhere over the last few decades. Well, almost everywhere. There are a few experiences that you've been carefully rationing for yourself.

The time has come for one of those trips.

With Pluto waking up as it gets close to perihelion—the closest point to the Sun in its orbit—the atmospheric pressure has now risen to the point where you can see the layers of tholin in the atmosphere as those robotic explorers of old once did. But this is the first time it has been seen by human eyes—your eyes.

To heighten the experience, you need the right vantage point. You can think of none better than peak T2 in the Tenzing Montes. Rising over six kilometers above the nitrogen ice plain of Sputnik Planum, this is the tallest mountain on Pluto. By prominence—a mountaineering measure that takes into account not just the altitude of the peak but how high that peak is above its base—compared with planet size, T2 is ranked behind only Olympus Mons on Mars and the peaks of Io as the solar system's largest mountain.

Yet where Olympus Mons' volcanic shield is a gentle stroll up a slightly sloped plain, T2 looks more like an unclimbable alp. The geology of the feature is unique. All the peaks in the Tenzing range are made of water ice that appears to be "floating" in a sea of denser nitrogen ice. It is as if the impact that created Sputnik Planitia fractured a water ice crust and the pieces that were not obliterated or ejected into space were frozen in place like stone icebergs.

It also doesn't hurt that you can see T2 in what is, to you, the most iconic image ever taken of Pluto, perhaps in all planetary science. As the New Horizons spacecraft departed, it snapped an image looking back at the limb of the planet along the terminator. There you can see long shadows cast by the Tenzing range and even some cold clouds amid the peaks. The vantage point reminds you of the romantic painting *Wanderer Above the Sea of Fog* by Caspar David Friedrich.

All that was missing in that image was the observer—though you share very little with the protagonist of Friedrich's painting, perhaps that could be you this time around the sun? A friend of yours certainly thinks so—they're coming to visit and plan to write an article on you and your research. They're going to take their ship to meet you on top of T2 for the best imagery to accompany their writing. You can't deny that you're looking forward to the company!

But that's the easy way up. You know you'll enjoy the experience much more if you must work for it. So rather than descend on a rocket to that elevated place, you're climbing up from the plain below. For weeks, you've been ferrying gear and supplies out to the mountain and you've been laying ropes and caches. As the time for the interview draws near, you feel you're ready to make the ascent.

At last, the day is here. You've got this timed out not just to make the meeting at the summit, but also because you want the lighting to be just right. Pluto's day lasts just over 153 hours, so sunset will happen slowly. Still, you give yourself a solid 24 hours of margin just in case. Certainly, it won't hurt to spend a little longer at altitude admiring the view.

The hike up to the Tenzing Montes is certainly dramatic. You use your vehicle to get as far as the edge of the dune fields that surround the mountains. From here, the great icebergs of the range already loom as the Sun just peeks over the horizon. Stepping out with your pack, you navigate around pits in the nitrogen ice. The sublimation from these features is critical to lofting the raw materials needed for the dunes, and with Pluto near perihelion, you can see the process at work as you walk in puffs of glittering particles. There was some question when you arrived as to whether it was nitrogen or methane ice particles involved in forming those dunes. You chuckle: it was an easy but enjoyable experiment to sort that one out.

Next, you walk over the dunes themselves. Their undulating pattern rises to as much as 30 meters in some places, and you know from experience that you don't want to step in the wrong area. Some parts of the dunes are compacted, while others are loose, and it can be difficult to tell the difference. Luckily, the dunes don't move particularly quickly, and you have a well-worn pathway to the first water ice foothills. Within a few hours, you are there.

Now things become a bit more technical. You weave your way here and there between the blocks. In places, you've put in ropes to surmount an obstacle that would take you hours to circumnavigate. When you climb, it's not like icefall or glacier climbing at home; instead, at 40 K (−233°C) the water ice has the consistency of concrete. That means you can't easily chip in to make footholds and there's no point to ice axes, but at least your pitons stay put and there's no danger of icefall in the transitional hours to sunlight.

Within a few more hours you arrive at the first of several vertical faces. Time to have a meal and rest up. You pitch a pressurized tent to give you some relief from wearing your suit. Outside, the weather seems quite calm. The Sun is still up, and it will stay that way throughout the climb—your plan was to complete the entire ascent in 72 hours (or so) of sunlight. As for wind, it's hard to generate much force with just a few pascals of pressure, tens of thousands of times less than we experience on the Earth—in fact, the air passing over the tent barely makes a sound.

A solid eight hours later, you roll out of your tent and pack the whole thing up. From here on out, you'll be carrying your home with you as you climb like a hermit crab. But at a gravitational acceleration of just 0.063 $g$, it would take a lot of deadweight to slow you down. Looking back toward the plain, you start to see a bit of glittering haze. The Plutonian day progressed while you were asleep and the sublimation in the pits picked up, launching particles skyward and adding pressure to the atmosphere. You were right to make your walk over the dunes close to sunrise when the activity was at a minimum.

You waste no time and start in with your mechanical ascender where the slope leans backward and you've placed some fixed ropes. From here you still have over five kilometers upward to travel. Some of that ascent will be hike-able, but most will be a climb—T2 has an average slope of just under

20 degrees. You feel yourself working up a sweat. But before your body can get overheated, your suit's environmental system steps in to whisk the moisture away and to cool the air inside.

It's not just because you are completing this climb on Pluto that the experience is worlds away from what this range's namesake, Tenzing Norgay, would have encountered. You're not sure you'd want to take this on in full gravity, without the oxygen and thermal control provided by your suit. You've heard that some people climb high mountains on Earth without oxygen—but that wouldn't be your first choice for adding a degree of difficulty to your experience. It's easier to enjoy what you are doing when your brain's physiological needs are satisfied.

As you get higher, your view changes. The plain, far below, appears to recede and you start to appreciate the companion peaks to T2. You pause here and there to look at imperfections and interesting inclusions in the ice. A human being is prejudiced toward thinking that any ice should be completely clear and crystalline, since that's our experience of the stuff. But this ice is like rocks elsewhere, full of impurities that give rise to strange color variations and shapes. The translucency of the material just adds to its mystique, since you can see features below the surface that would otherwise be invisible to a climber. You're going to have to return later with some extra equipment to study this stuff in full! This ice might have once been well below the surface of Pluto, and now here it is, lying on the open ground.

With significant effort, you manage to reach nearly to the top of your fixed ropes by the end of your second day of climbing. You give thanks once more for the 0.063 $g$! From here it looks to be a relatively short traverse up to the summit, maybe a kilometer as the crow flies with the first half on fixed ropes. That's only about 10 percent of what you've already climbed. Still, your muscles are starting to complain, and you'll appreciate a hot meal and a rest. You've intentionally picked a bivouac dangling off the side of a sheer drop. It feels a bit irresponsible, but the tent is built for this, especially once you've added an extra couple of pitons to prevent any swaying. Before you take off your suit, you just sit there on the edge, feet hanging over open space, lost in thought.

The next morning, something seems off with the light. Once suited up, you unzip the flap of the tent and find yourself staring at a fuzzy wall! At first, you are confused—but then you realize that you're looking at ice fog—it's now early afternoon during Pluto's day and sublimation and winds are at their maximum, putting icy particles into the air. It's astounding that it could come up to this altitude! There's no danger to you from the particles themselves or from the light winds, but the reduced visibility is going to be a challenge.

You don't relish finding your last cache, perched high on the mountain. You set off along your fixed ropes in the eerie cloud. For the next hour it's just you and the mountain face, like some Platonic ideal of outer solar system climbing. Around lunch you reach the promontory where your cache should be, but you don't see it! Frantically, you start looking around—you'll need the oxygen you have stored soon enough. But the strange lighting is playing tricks with your vision, and there's a thin dusting of snow over everything.

Suddenly, your boot connects with something solid. At last, it's the cache! You sit down and take a rest, changing out your oxygen and replenishing your supplies. Even as you sit, the light seems to be changing, with the air above you becoming darker. Within minutes, the fog clears and falls below your level. You gaze across the open air and it's as if you and the other peaks, like islands, are alone on a fuzzy sea filling in the space between the peaks with a dark sky above. You could almost imagine sailing a boat across to the other parts of the range. It's a strange, ethereal sight.

But you've lost time and now you intend to make it up. You reach the top of the fixed rope easily and then you apply your focus. For the rest of this third day (and much of the remainder of Pluto's daylight) you pick your way up the remainder of the mountain. This section is considerably less steep than you had feared, and you make good time.

It seems as if the ascent will never end, with false summit after false summit, and then, suddenly, surprisingly, you are there! You look around, taking it all in from horizon to horizon. Off to one side, the Sun is getting low. Below, the fog has been surging up and down all day and now seems to thin out and rise as if dispersing. Some of those particles make it all the way up to your level.

As they do, a dark, vaguely human figure seems to materialize in front of you, opposite the Sun, surrounded by a many-colored halo. Impossibly large, the figure seems to cut through the thin fog as far as you can see. You move your arm and discover that the figure is a projection of yourself, an optical effect known as the "specter of the Brocken." Well, how about that. Sounds like a good topic for a paper, you muse.

Eventually, the figure disappears, and the fog thins out. Behind, you start to see a brightening star in the sky. It's the dim fire of thrusters—your friend will be arriving soon. As marvelous and spellbinding as this vista is, it's something that's better shared.

The small lander touches down and your friend steps out and over to you. They make as if to wave but stop. Behind their faceplate you can make out an awed expression. You've become inured to Pluto's wonders these past years, but what must this be like for a first-time visitor? Your friend is speechless; their camera hangs heavy on its straps, forgotten in the moment.

They come over and put a hand on your shoulder before you both turn to admire the view. Together, you stare out at Sputnik Planitia with eagle eyes, surveying the scene and looking at each other with wild surmise, silent upon this peak in Tenzing.

★★

This chapter is entitled "Pluto, the Edge of the Map," but perhaps that may feel a bit misleading now, doesn't it? In chapter 9, we looked at the highly eccentric orbits of comets, and dissected the difference between the Kuiper Belt and the Oort Cloud. Knowing that, we realize Pluto is really only a small step out into our solar system. But in terms of what we have learned about this system of planets, dwarf planets, moons, and more, Pluto really does represent the edge of our knowledge.

Back in July 2015, the spacecraft New Horizons made a flyby of Pluto when it was about 34 au from the Sun, revealing a complicated, dynamic, and beautiful world that we were not expecting.[1] The data sent back by this spacecraft wholly and forever changed our understanding of the planet-become-dwarf-planet that everyone loves, and basically rewrote every textbook on the subject.

But first, a quick note: the New Horizons spacecraft performed a flyby mission of Pluto, which means it didn't hang around; it just flew right by heading farther out into the Kuiper Belt, in which Pluto is found. While its primary mission at Pluto was complete, mission scientists searched for other possible Kuiper Belt objects that happened to be along New Horizon's flight path beyond Pluto and narrowed them down to one. In January 2019, New Horizons flew by its second target, Arrokoth, at 44 au from the Sun, making it the farthest and most primitive object we have ever visited.[2]

As of early 2024, New Horizons is 58 au from the Sun, and continues to travel farther. The team is hard at work searching for new potential targets they may encounter along the way, as New Horizons heads out of the solar system forever.[3]

## A NOTE ON PLANETARY DEFINITIONS

As we briefly mentioned in the endnotes of chapters 4 and 10, there is an official definition of what is considered a planet in our solar system. This definition was created and voted on in August 2006, at a meeting of the International Astronomical Union, the international professional society of astronomers. As it turns out, when New Horizons launched in January 2006, leaving Earth and heading for the edge of the map, its target, Pluto, was considered a planet officially. But when it arrived in 2015, Pluto had been reclassified to dwarf planet.

The official definition of a planet requires three criteria to be met:[4]

1. The object must orbit a star.
2. The object must be large enough and massive enough to have crushed itself into a sphere.
3. The object must be gravitationally dominant enough in its orbit to have effectively cleared it out or otherwise shepherded surrounding debris into stable resonances.

Pluto meets the first two criteria but fails the third. Since Pluto is part of the Kuiper Belt, it shares its orbit with many thousands of small, icy objects (some of which are or will become short-period comets!) over which it has absolutely no control. Therefore, Pluto did not clear its orbit, so it is not considered a planet. However, the new category of "dwarf planet" was created to account for objects that meet the first two criteria but not the third. Currently, there are five official dwarf plants in our solar system: Ceres, Pluto, Eris, Haumea, and Makemake.[5]

The decision to reclassify Pluto remains controversial to this day, even within the astronomical and planetary science communities. There are many prominent scientists who believe Pluto should still be considered a planet (along with the other four dwarf planets!), and many others who make a good case for the opposite. But regardless of your stance on the current definition of Pluto, if we take a step back and look at the context here, what this debate represents is scientific progress.

The whole reason this debate began in the first place was because we started finding objects that share the same orbit as Pluto. We can't ignore this data. As scientists we either must (1) find a way to fit those new data points into our current paradigms of understanding or (2) smash the models and make a new one that fits all the data better.

With Pluto, data was mounting for years that many felt was demanding the second route. In a very real sense, we are watching the scientific process unfold. And that is pretty cool.

As a group, scientists like to debate, challenge, prove each other wrong, and, ultimately, find a system of knowledge that approximates nature as best as possible. But we aren't perfect. Maybe this definition is wrong. New data may be found that challenges this current definition and demands smashing of the models again. In fact, this may be already happening with the exponential growth in discovery of exoplanets[6] (see the epilogue as well).

Here's one final note: this debate about Pluto is eerily similar to another debate that occurred in the astronomical community regarding the discovery of objects in the gap between Mars and Jupiter. Back in 1801, the object Ceres was discovered. Originally, it was called a planet, but more objects were found in the following years and decades. This led to the discovery, naming, and classification of the asteroid belt.[7] Ceres has gone from planet to asteroid to dwarf planet in the last 200 years.

### THE MAGNITUDE OF A COLD AND DISTANT SUN

It is often said that the Sun is just another star from the perspective of Pluto, but would that really be the case? To answer that question, we can compare the brightness of the Sun from the distance of Pluto with that of the brightest star in the sky as seen from our solar system, Sirius. The best way to compare these two stars is to use a scheme called *magnitude*, which describes how astronomers calculate brightness.

The magnitude system can be traced back to an ancient astronomer named Hipparchus (c. 150 BCE). Hipparchus created a catalogue of stars, and grouped all the stars he could see into six brightness groups. The brightest stars were considered first-magnitude and the dimmest were considered sixth-magnitude. Of course, this does not allow for much nuance and, most importantly, did not account for stars too faint to see with the human eye. Today, astronomers still use the magnitude system created by Hipparchus, but have expanded it to include dimmer stars and brighter objects as well (such as the Sun and Moon).[8] As a result of keeping Hipparchus' convention, the lower the magnitude number, the brighter that object is. For example, a star with magnitude 1.0 is brighter than a star with magnitude 3.0. Particularly bright objects even have negative numbers.

The magnitude scale is also logarithmic. Specifically, the modern version of this system was designed to have a brightness ratio of 100 to 1 from a magnitude of 1.0 to 6.0. That means each step in magnitude is equal to a difference in brightness of about 2.5. But it turns out that this scheme is exactly how our eyes perceive the universe, making it a surprisingly intuitive scale for us. This makes sense, as it was originally conceived directly from naked-eye observing. Of course, this is a difficult scale for digital cameras, like the ones used by our spacecraft. This is why you must be very careful to set the exposure of a camera properly, but your eyes don't need the same kind of adjustment.

In the magnitude scale, Sirius has a magnitude of −1.46, which is the same whether you are on Earth or Pluto because Sirius is so far away. By contrast, the Sun, as seen from Earth, has a magnitude of −26.74. Because Pluto is so much farther from the Sun than is the Earth, and the brightness of the Sun diminishes with the square of the distance,[9] we need to take Pluto's orbital distance into account.

Like any orbit in the solar system, Pluto's is an ellipse; however, Pluto's orbit is very eccentric, with a measured eccentricity of 0.24.[10] At its closest approach, it is just a bit over 30.1 au from the Sun. At its furthest, Pluto is just over 48 au away. As such, the apparent magnitude of the Sun, as seen from Pluto, varies between −19.3 and −18.3. This is still far brighter than Sirius.[11]

Even way out at the edge of the map, the Sun is still able to deliver enough energy to give Pluto seasons, though they have more to do with the change in distance from Pluto's eccentric orbit than with the axial tilt, which drives the seasons on Earth. When Pluto is at perihelion, the atmospheric pressure rises tremendously, and lots of atmospheric processes, such as those described in the story, take place. Once past this perihelion point, Pluto goes back into the metaphorical cooler. The atmosphere condenses out on the surface, and everything returns to a stasis-like mode until the next perihelion passage. It just so happened that Pluto recently completed a perihelion in 1989. With an orbit that takes about 248 years to go around the Sun once, Pluto won't be back at perihelion until 2237.

## TOBOGGANING ON ANOTHER WORLD

Despite the unearthly setting, the physics of tobogganing is no different on Pluto than it is on Earth. Though the gravity and therefore the acceleration is much less on Pluto, any low-friction material placed on an inclined plane will accelerate down the slope and can pick up significant speed if that acceleration goes on long enough. On Pluto, typical methane ridges (or dorsa) are about 500 meters high, separated by about three kilometers, and sloped about 20 degrees to the horizontal on average—a considerable incline for tobogganing.

That geometry gives us an effective acceleration of 21.5 cm/s$^2$ down the slope (in the absence of any friction). With about 1,500 meters of downslope run, at that speed it will take just under two minutes to get all the way down, and our astronaut would be traveling at about 90 km/h by the time they arrived at the bottom. As they rise up the opposite slope, gravity would begin to slow them down until they arrive at a gentle stop. By the way, you can measure the friction in the system between the plastic and the methane snow by observing how high up the opposite ridge the toboggan travels. It's likely our protagonist has tried out this experiment!

## PLANETARY EVOLUTION

The giant iceberg mountains discussed in the story are unique to Pluto, but they reveal an interesting planetary process.[12] Planets (and dwarf planets) are built by many smaller objects the size of asteroids or comets, or maybe even larger, smashing together, one by one, over many millions of years. Slowly, the objects accrete onto the newly forming planet. All these objects that built the planets were made of a variety of different materials with different densities; they would have had a variety of metals, rocks, and ices all mixed very thoroughly. Yet, as we discussed in chapter 4, large objects end up differentiated, with layers of metal (the core) and rock (the mantle and crust) for the Earth, for example, and layers of rock (the core) and various ices (the mantle and crust) for Pluto. How do the solid objects differentiate?

The answer is heat. Initially, this heat comes from the energy of each new piece of material that strikes the growing planet at high speed, just like any other bolide. The kinetic energy is turned into thermal energy, which melts the planetary crust, allowing what is left of the impactor and the nearby crust to separate out into different components of different density. But there's a problem with this "heat of accretion": it is deposited right at the surface, and therefore the hot crust can radiate that heat away into space. It doesn't penetrate very easily into the interior.

Luckily, there is another source of heat: radioactive decay. Today, the same process is responsible for powering our fission reactors, which heat up water into steam and turn turbines to create electricity. Early in the solar system's history, the concentration of radioactive elements was much higher than it is today. These radioactive elements were able to heat up the interiors of planets more thoroughly, allowing the melting needed for dense materials to separate from less dense materials. Once the process gets started, the friction from blebs of planet pushing past one another like in a lavalamp generates even more heat! The densest materials, such as iron, descended into the Earth's core. Denser rocks, such as olivines, ended up in the mantle, and the lightest rocks, such as silicates and basalts, ended up in the crust. The same thing happened on Pluto, with rocks in the core, denser ices like nitrogen ice in the mantle, and less dense ices like water ice on the crust. See figure 15.1.

And this brings us to one of the most iconic features on the surface of Pluto: officially named Sputnik Planum but known simply as Pluto's Heart.[13] In the context of this discussion, the curious thing about Pluto's Heart is that it is made of nitrogen right at the surface—how could this happen? It seems likely that at some point in the past, a large impact struck Pluto, removing much of the water ice from Sputnik Planum and allowing some of the nitrogen ice to rise to the surface. With the energy delivered by the impact, there was likely liquid present for a limited amount of time. At the periphery of the impact, the crust would not have been vaporized or ejected, but simply broken up, and these gargantuan icebergs of water ice crust floated at the edge of this new ocean. But this state of affairs couldn't last. Eventually, the transitional ocean froze solid, and the water ice chunks were left in place to stand as mountains.

This impact likely influenced the whole of Pluto. Sputnik Planitia is located on the equator almost exactly opposite Charon, Pluto's tidally

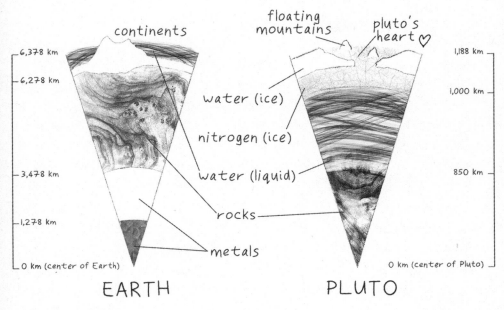

Figure 15.1
In comparing the interiors of the Earth and Pluto, many resemblances can be seen.
The continents of the Earth and iceberg mountains of Pluto are fascinatingly similar.

locked satellite. If the impact that created this plain brought enough dense
nitrogen ice to the site of the impact, it could have created an especially
large concentration of mass at the latitude and longitude of Sputnik Planitia. Since objects that spin, like planets, tend to spin most stably with the
largest concentration of mass at the equator, this may have reoriented the
poles of Pluto over time. Mars likely experienced a similar process as volcanoes built the massive Tharsis plateau.

# EPILOGUE: INFINITE DIVERSITY IN INFINITE COMBINATIONS

Your sleep was fitful last night after the journey up to orbit. Even without pressure points, you find that you can never quite get comfortable trying to doze off in zero gravity. Before you could fully submerge yourself in sleep, dreams of falling through cloud decks in deep gas giant atmospheres resurfaced you to waking.

It's just as well—today will be an early day. You have a flight to catch that will take you farther than you have ever traveled. You're looking forward to diving into deeper waters than you can find in the neighborhood that you have explored thus far.

Following a quick shower, you gather your personal effects into a nylon carry-on bag. Everything else you need is already loaded onto your ship. If only you had this kind of service when you were ballooning on Venus! But there are a few personal mementos that you want to keep with you. A dusty collapsible golf club. A jar containing a mixture of water formed from melted ice samples you collected from across the solar system. An overbuilt rock hammer and a length of slender electrostatic rope.

You pause to make a cup of coffee by injecting hot water from a dispenser into a heatproof plastic pouch with freeze-dried grounds, sugar, and cream. After kneading the bag to ensure everything is dissolved, you carefully squeeze the hot liquid into your mouth. It does the job. But somehow, it's just not the experience you were anticipating. Maybe if you could add just a little gravity to the mix, or better yet could ramp up the air pressure to 50 percent above Earth normal.

Taking the pouch and the carry-on with you, you exit your hotel room and take a glance back at the poster of Saturn's moon Hyperion on the wall before closing and locking the door. Floating along the corridor through the cave-like space, you catch glimpses of different views of the outside through wide portholes. Out one, there is the gleaming marble of a planet turning slowly below you. Another is a window to a glittering starfield—diamonds on black velvet.

But the one that interests you the most right now is coming up next. Looking upward from your perspective, your ship looms overhead like a submarine tethered to a gantry in an endless dark ocean. It's larger than any you've seen before. That's because it has farther to go. Your journey isn't to Jupiter or Saturn or even to Pluto. No, you're heading out into the deep black to cast your eyes on planets around other stars. Places that have never basked in the light of our own Sun.

These extrasolar planets, or exoplanets as they are known, can be a lot like home. But they can also be strange places. Some are located so close to their parent star that their surfaces and atmospheres are superheated. These lava worlds are locked in place, with one side of the planet molten and the other side bathed in rocky rains.

Other solar systems feature gas giants so large they make Jupiter look small, or have carbon as their primary solid-forming element instead of the oxygen that makes up the bulk of our silicate rocks and icy moons. Some planets revolve around smaller, colder, redder dwarf stars that will keep on burning long after our Sun has become a white cinder. Others wander the cold empty spaces between the stars, rogues beholden to no particular gravity well. No doubt there are worlds of endless oceans and others with biologies that we can scarcely imagine. Or planets located in star systems with several suns. Truly, there are more things in the heavens than can be dreamt of in any philosophy!

Most intriguing are the super-Earths. These especially common planets are bigger than the Earth but smaller than Uranus or Neptune, the smallest gas giants in our solar system. This means that they still ought to be dominated by rocks with a solid surface deep down somewhere but are swaddled in far more gas than any planet in our solar system. With that extra material between their surfaces and space should come extra protection, something that might make these places better homes for life than even a copy of our own home world elsewhere in the cosmos.

But we don't know! Because we don't have an example in our own solar system, the only way to find out for certain is to investigate examples of places like these in other solar systems. Having all these strange new worlds means that, for the first time, we can really understand the general processes that lead to different planetary environments. Taking the census of the galaxy's hundred million–plus exoplanet representatives reveals the astounding diversity that simple physical processes can create.

And you are headed out to see a handful of them.

What would it feel like, you wonder, to have the wind in your hair as you sail on a world-wide ocean? To visit a place where you might dive onto coral reefs populating the highest sunken mountains? Or to hike around the circular stationary terminator of a lava world, taking time only to lay back and watch the clouds and optical displays formed as rock vapor condenses into crystal snow? Could anything you experienced hiking on Mercury even begin to prepare you for that experience?

Luckily, you will have the chance to find out.

You reach a junction tube and change your direction to propel yourself toward the ship. Before entering, you finish your coffee and toss the empty pouch into a nearby receptacle. When you are traveling across the light years, any saved mass helps!

The interior of the ship looks much like the hotel and the space dock. A simple repetition in functional design. After stowing your carry-on in your quarters, you head up to the control room. You say hello to the captain, an old friend, and then take up your station. There are no idle passengers on this ship, not even you: everyone helps in ways large and small that weave the fabric of the crew together. Besides, you have some navigation experience from your time out at Saturn that is just as valuable in an interstellar cruise as it was chasing comets. Out in the galaxy, as in the solar system, Newton's and Einstein's laws still apply.

You take a second to clip a photo to the corner of your console. The captain, an avid photographer and former journalist, looks over and offers you an appreciative smile. The image background is a symphony of jagged blacks and grays. But shining in the middle is a proud splash of color encased in a bell jar. You lean back and contemplate the rose as you begin your next journey and daydream of worlds yet to come.

# AFTERWORD: WE ARE ALL EXPLORERS

In the early era of the exploration of space, eligibility to travel beyond Earth was restricted to a very small number of people from a remarkably limited range of backgrounds. To give just one example, the first astronauts of the US National Aeronautics and Space Administration, the Mercury Seven, were all white, male, five-foot-eleven or shorter Christians from American towns with military backgrounds. Yet the human desire and capacity to explore and experience knows no gender, ethnic, religious, racial, geographical, height, or any other boundaries.

This is where our own internal biases can deceive us. If we were to describe someone to you as an astronaut or an explorer, the image that forms even within your own mind might look a lot like one of those Mercury Seven. You might even think that "explorer" or "astronaut" are categories that cannot possibly include yourself. But nothing could be further from the truth. All of us are here today because of the explorers who came before us. The Polynesians who set out across the ocean searching for new lands not knowing whether they would find an island at the end of their voyage were explorers. No less were the Indigenous peoples who crossed the Bering Sea land bridge during the last glacial maximum. Any close reading of history and archaeology will find that all cultures are explorers. Therefore, if you are human, you are already an explorer.

Indeed, as any new parent with a toddler will attest, there is a need to explore one's environment that is programmed very deeply within each one of us.

That exploration does not need to center on risk to be genuine, and it's not just about geographic travel. Athletes explore the possibilities of human physical achievement. Others explore the limits of the possible within our universe through science, engineering, or art. For whatever reason, one of the factors that makes us distinctly human is our curiosity and our need to make the unknown known. Sometimes that means getting up and going out your door. But at other times that voice inside takes us down paths that are metaphorical.

When we wrote the stories that appear in this book, it was our greatest desire for you to be able to close your eyes and imagine yourself in the shoes of the protagonist, no matter who you are. For that reason, we were careful to include as few details as possible on the identity of the protagonist, to break down the barriers between author, subject, and reader as much as possible, given our own unconscious biases. If we did our job well, you might have been able to tap into that feeling of being there, just a little bit.

As a final note, in writing a book describing exploration, it was necessary to consider the relationship between the protagonist and their environment. There are two end-member perspectives that are possible here. One can take a colonist perspective, in which the planets are simply resources to be exploited, and what matters is what can be achieved with those resources by human hands. At the other end of the spectrum is an ecological perspective in which the planets have inherent value just as they are.

Early work in science fiction tended to look toward the era of European colonization following the Renaissance for inspiration or, later, to the westward expansion of US settlers during the eighteenth and nineteenth centuries. This can be seen in work as recent as *Star Trek*, in which a remarkably diverse crew (for the 1960s) spends a fair bit of time either "planting" or otherwise checking up on the Earth's colonies around other stars. This is unsurprising—at the time, Westerns were a popular genre and the TV series was sold to the network as "*Wagon Train* to the stars," a reference to a popular genre show of the period. To its credit, *Star Trek* did, and not just occasionally, aim to subvert that same genre by seriously treating the unintended consequences and moral ambiguity of that colonist perspective.

More recent work has taken the dichotomy head-on, wrestling with the tension within and between each perspective. Perhaps the most explicit consideration of that conflict lies within Kim Stanley Robinson's Mars trilogy. The three books (*Red Mars*, *Green Mars*, and *Blue Mars*) describe the explorers as being divided into factions between those who look to Mars and see great works of transformation to be accomplished and those who would make it a wild park, frozen in time, preserved forever. Of particular interest is the idea of an ecological perspective even in the absence of an ecology!

The conclusion of the *Mars* novels seems to be that both perspectives have value and that both are necessary for something truly unique and beautiful to emerge, that we too are part of the environment, especially on other planets that may—unlike other lands on the Earth—truly be empty and sterile.

We have taken a similar tack, perhaps one even more extreme than the Mars novels in that we are describing environments much as they are today, typically with very few human touches. Where human infrastructure is present to allow a story to plausibly take place, the protagonist often stays as far in the wild as possible. A surprising aspect of those stories was that they required us to explore the conflict between perspectives, with the protagonist often taking the ecological side, in contrast to those who sought to use the environment for their own ends. As the painter Bob Ross once said, to get the full range of hues, "You absolutely have to have dark in order to have light."[1]

In the end, space exploration is utterly dependent on technology and society. Space is an inherently lethal place for an unprotected human being. That necessarily means that we will bring along the necessary infrastructure to support not just ourselves, but also our machines. As such, living lightly on the land on Titan will look very different from trying to do the same in a tropical location on Earth.

Given this constraint, why not just leave these places alone?

Well, we are convinced that there is value in travel and in experiencing these places, that broadening one's horizons allows us to see our own home and those who inhabit it with new eyes. It also gives that place life within human culture as we visit such locations and then relate the experience to others. The Apollo astronauts are living examples of such storytellers, many of whom felt compelled to share their experience of the magnificent desolation they encountered on the lunar surface. To have knowledge of such places allows us to include them in our worldview and to tailor our actions with them in mind. This prevents us from viewing these places simply as blank canvases on which to paint future versions of our past mistakes. Instead, we can see them for what they truly are and can imagine ways in which we can live in harmony with the many worlds of our universe.

Through our stories, we hope that we have allowed these places to live in your imagination. No doubt we have erred in those places where we filled in the gaps between what is currently known and what can only be imagined. In these gaps lies an even more beautiful truth, waiting to be discovered. In the future, we hope that many of you or your descendants will be able to travel and explore in order to experience the stories of these places and to retell their stories to others.

## ACKNOWLEDGMENTS

We hope you have enjoyed our visits to these alien landscapes. And if you've come this far, perhaps you'll come a little further.

This book arose out of something that we couldn't find out in the world—a book that used what we know from scientific exploration of the solar system to inform the worlds we have created with our imagination. Such treatments exist commonly in science fiction (with some creative license) but have little or are completely lacking in science fact, which keeps the reader at a remove. John has been involved with several space missions and has always deeply enjoyed the exploration aspects of his work. There is something magical about being among the first to view a new place, previously unseen by human eyes.

We are indebted to many in our journey, not least of which to the supervisors who trained us and taught us to read the maps of other worlds. For Jesse, that would be Patrick Hall at York University, and for John, Peter Smith of the University of Arizona. Furthermore, we could not have sustained the work without the support of our families and our partners Alex and Michelle (the latter would later go on to become our illustrator!). Our big break was when one of our academic colleagues and friends, Ray Jayawardhana, introduced us to the MIT Press. That path led us to our editor, Jermey Matthews, who immediately grasped the method in the madness of what we hoped to accomplish and who became a critical advocate for the book you now hold in your hands. Once Jermey left the press, it was Haley Biermann and the fantastic crew at MIT Press who took up the torch on his behalf and helped us to get our text over the finish line.

Along the way, others helped us to refine our writing and the diagrams. We are indebted to our reviewers: Haley Sapers, Paul Delaney, Alasdair Petersen, Dan Riskin, and Cassandra Marion, who provided their time, their feedback, and their advice completely free of charge. Without them, we would not have been able to express our ideas with such clarity and impact. Moreover, we are delighted to have been able to involve them all in this work during its infant stage so that they could share just a little bit of the journey with us.

Speaking of infants, this work has truly been a labor of love. From a concept that Jesse and John put together while John was on sabbatical in 2019, this book was nearly five years in the making. During this period, the structure of the text morphed several times before settling into the format you find before you. When you work on something for that long, you become aware of the passage of time. We feel particularly fortunate that we were able to complete all the stories in time for our first reader, Elaine—John's mother—to listen to them being read aloud during her final week before passing in April 2023.

But it wasn't all departures and losses—there were also joyful arrivals. In August 2021, Alex and Jesse welcomed their daughter Charlie into the world. We look forward to sharing these stories with her.

Finally, both Jesse and John would like to acknowledge Robert J. Sawyer, our foreword writer. He is among many science fiction authors who have inspired and influenced our work, and we have always appreciated the realism of his writing, which seamlessly and impactfully stitches together the present and the (near) future. That he does it all with a Canadian angle really hits home for us.

# NOTES

## CHAPTER 1

1. Dana Bolles, "The Moon," edited by SMD Content Editors, NASA, accessed December 29, 2023, https://science.nasa.gov/moon/.

2. Gravity also accelerates electromagnetic radiation, aka "light." While light does not have mass, it does respond to curved space-time. This may not make much intuitive sense, but you can find more information at https://www.science.org.au /curious/space-time/gravity.

3. However, due to the drag created by Earth's atmosphere, there is a limit to the speed a falling object can reach. We call this "terminal velocity." If Earth had no atmosphere, though, any falling object would continually accelerate until it hit the ground. We'll come back to this concept in chapter 11 when we go skydiving in Jupiter's atmosphere.

4. Ten meters per second is about 35 km/h. So imagine accelerating from 0 km/h to 35 km/h in one second, then to 70 km/h the next second, and so on.

5. Earth isn't a perfect sphere. If you measure its radius from center to pole, you get 6,356 km. If you measure its radius from center to equator, you get 6,378 km. In other words, Earth is ever so slightly flattened. David R. Williams, "Earth Fact Sheet," NASA, June 5, 2023, https://nssdc.gsfc.nasa.gov/planetary/factsheet /earthfact.html.

6. In contrast to the Earth, the Moon is almost a perfect sphere, differing by only a couple kilometers in polar and equatorial directions. David R. Williams, "Moon Fact Sheet," NASA, December 20, 2021, https://nssdc.gsfc.nasa.gov/planetary /factsheet/moonfact.html.

7. To see just how bad light pollution can be, and where the best/worst locations are, see the interactive map at https://www.lightpollutionmap.info/.

8. That's right, stars twinkle because of the turbulence in our atmosphere. That means that astronauts above the atmosphere, like aboard the International Space Station, don't see the stars twinkle.

9. The prefixes *peri* and *apo* come from Latin and refer to "near" and "far," respectively. These prefixes are added to terms that refer to the object being orbited. In

the case of Earth "gee" is used (like geo), in the case of the Sun "helion" is used. Thus, for an object orbiting the Sun, what is the point in its orbit where it is closest called? What about the farthest? Answers can be found in the endnotes of chapter 4 and in the text of chapter 9.

10.  Paul Chodas, ed., "Glossary," Center for Near Earth Object Studies, NASA, https://cneos.jpl.nasa.gov/glossary/au.html.

## CHAPTER 2

1.  Dana Bolles, "Mars," edited by SMD Content Editors, NASA, accessed December 29, 2023, https://science.nasa.gov/mars/.

2.  The tallest peak on Mars is Olympus Mons, standing 22 km above the datum, and the lowest place is in Hellas Basin, which sits about 7 km below the datum. "Atmosphere," Mars Education, accessed August 25, 2023, https://marsed.asu.edu /mep/atmosphere.

3.  One hundred megapascals is about 1,000 times Earth's atmospheric pressure at sea level.

4.  Pippa Whitehouse, "Postglacial Rebound," AntarcticGlaciers.org, June 22, 2020, https://www.antarcticglaciers.org/glaciers-and-climate/sea-level-rise-2/recover ing-from-an-ice-age/.

5.  David R. Williams, "Mars Fact Sheet," NASA, May 22, 2023, https://nssdc .gsfc.nasa.gov/planetary/factsheet/marsfact.html.

6.  This is a topographic map of Mars. Data was collected by NASA's Mars Global Surveyor and is presented in an easy-to-use map by Google. Can you find Olympus Mons? Where's Valles Marineris? What about Hellas basin? See https://www .google.com/mars/.

7.  "Highlights of Canada's Geography," Statistics Canada, October 7, 2016, https://www150.statcan.gc.ca/n1/pub/11-402-x/2012000/chap/geo/geo-eng.htm.

8.  Daniel Stolte, "The Reason for Mars' Tumultuous Past," *University of Arizona News*, April 22, 2016, https://news.arizona.edu/story/reason-mars-tumultuous-past.

9.  Walt Feimer, "Mars' Ancient Ocean," Scientific Visualization Studio, May 3, 2023, https://svs.gsfc.nasa.gov/11796.

10.  "Cracks in Ancient Martian Mud Surprise NASA's Curiosity Rover Team," NASA Science Mars Exploration, August 9, 2023, https://mars.nasa.gov/news /9459/cracks-in-ancient-martian-mud-surprise-nasas-curiosity-rover-team/.

## CHAPTER 3

1.  Dana Bolles, "Europa," edited by SMD Content Editors, NASA, accessed December 29, 2023, https://science.nasa.gov/europa/.

2. Jupiter has 95 natural satellites, as of May 2023. Check out the current list here: https://nssdc.gsfc.nasa.gov/planetary/factsheet/joviansatfact.html.

3. Back in the early 1600s, it still wasn't fully accepted that the Moon was a "natural satellite" of Earth. Many believed that the Earth was the center of the universe, and the Moon orbited us, just like everything else. It wasn't until Galileo discovered the four largest moons of Jupiter with one of the first telescopes that others started to accept that planets can move through space and bring with them their natural satellites.

4. David R. Williams, ed., "Jupiter Fact Sheet," NASA, May 22, 2023, https://nssdc.gsfc.nasa.gov/planetary/factsheet/jupiterfact.html.

5. "Galileo Evidence Points to Possible Water World under Europa's Icy Crust," Jet Propulsion Laboratory, April 25, 2000, https://www.jpl.nasa.gov/news/galileo-evidence-points-to-possible-water-world-under-europas-icy-crust.

6. For a full overview of the Galileo mission, including its top ten most interesting discoveries, see Dana Bolles, "Galileo," edited by SMD Content Editors, NASA. Accessed December 31, 2023, https://science.nasa.gov/mission/galileo/.

7. Katrina Jackson, "Hubble Directly Images Possible Plumes on Europa," Scientific Visualization Studio, September 26, 2016, https://svs.gsfc.nasa.gov/12375.

8. "NASA's Europa Clipper," NASA, June 3, 2014, https://europa.nasa.gov/.

9. "From Soup to Cells: The Origin of Life," Understanding Evolution, November 17, 2021, https://evolution.berkeley.edu/from-soup-to-cells-the-origin-of-life/where-did-life-originate/.

CHAPTER 4

1. Dana Bolles, "Mercury," edited by SMD Content Editors, NASA, accessed January 2, 2023. https://science.nasa.gov/mercury/.

2. Terrestrial translates to "of or relating to Earth" from Latin. In effect, it means Earth-like.

3. Unless you include Pluto; some still do. Officially, Pluto is a dwarf planet, as decided by the International Astronomical Union in 2006, though there are many prominent planetary scientists who disagree with this decision. "Pluto and the Developing Landscape of Our Solar System," https://www.iau.org/public/themes/pluto/.

4. Of note, Mercury is only slightly larger than our own Moon, which is 3,474 km wide. Given the sizes of Mercury, the Moon, and Pluto, it makes one wonder if size should be considered a determining factor in planetary definitions. David R. Williams, ed., "Mercury Fact Sheet," NASA, May 22, 2023, https://nssdc.gsfc.nasa.gov/planetary/factsheet/mercuryfact.html.

5. "Running Up That Hill," NASA, November 22, 2014, https://photojournal.jpl.nasa.gov/catalog/PIA16536.

6. A solar day on Earth is 24 hours, everyone knows that one. But a sidereal day is 4 minutes shorter, at 23 hours and 56 minutes. See figure 4.10 of Openstax, Astronomy 2e, https://openstax.org/books/astronomy-2e/pages/4-3-keeping-time.

7. "Terry's Story," Terry Fox Foundation, October 6, 2022, https://terryfox.org/terrys-story/.

8. Frederick Dreier, "This Hiker Just Completed the 6,800-Mile Great Western Loop in Less Than 200 Days," Outside Online, November 15, 2022, https://www.outsideonline.com/outdoor-adventure/hiking-and-backpacking/nick-gagnon-great-western-loop-record/.

9. It took a while for humanity to let go of the idea that orbits in the solar system must be perfect circles. It took the work of a brilliant observer, Tycho Brahe, and the steadfastness of a truly dedicated mathematician to unravel the true nature of orbits in the solar system. Learn more about ellipses, eccentricity, and Kepler's laws of planetary motion here https://solarsystem.nasa.gov/resources/310/orbits-and-keplers-laws/.

10. Mercury changes its distance from the Sun from 46 million km at closest approach, a.k.a. *perihelion*, to 70 million km at its farthest, a.k.a. *aphelion*.

CHAPTER 5

1. David R. Williams, ed., "Venus Fact Sheet," NASA, May 22, 2023, https://nssdc.gsfc.nasa.gov/planetary/factsheet/venusfact.html.

2. Dana Bolles, "Venus," edited by SMD Content Editors, NASA, accessed January 2, 2023, https://science.nasa.gov/venus/.

3. Venera 7 was the first mission from any nation to land softly on another planet. That is to say, it touched down without crashing. https://nssdc.gsfc.nasa.gov/nmc/spacecraft/display.action?id=1970-060A.

4. For a brief mission profile of Venera 8, see https://nssdc.gsfc.nasa.gov/nmc/spacecraft/display.action?id=1972-021A.

5. For a great primer on what the ideal gas law is, see "The Ideal Gas Law: Crash Course Chemistry #12," Crash Course, May 7, 2013, https://www.youtube.com/watch?v=BxUS1K7xu30&ab_channel=CrashCourse.

6. This version of the ideal gas law, $P = \rho R^{*} T$, has a "mass basis," as opposed to the more well-known version, $PV = nRT$, which has a "number basis." In the former, the law is expressed in a way that is specific to a particular gas, since different gases have different masses, whereas $PV = nRT$ is based on the number of molecules in a given volume, which is more universal. It is possible to convert between these two versions of the ideal gas law.

7. The drag equation plays a large role in calculations related to rocketry. NASA has a brief description of it at https://www.grc.nasa.gov/www/k-12/rocket/drageq.html.

8. NASA's Magellan mission provided the first full RADAR map of Venus. See Dana Bolles, "Magellan," edited by SMD Content Editors, NASA, accessed January 2, 2023, https://science.nasa.gov/mission/magellan/.

9. Carolyn Jones Otten, "'Heavy Metal' Snow on Venus Is Lead Sulfide," The Source, February 10, 2004, https://source.wustl.edu/2004/02/heavy-metal-snow-on-venus-is-lead-sulfide/.

10. Dana Bolles, "Pioneer Venus Orbiter Map of Venus," edited by SMD Content Editors, NASA, accessed January 3, 2024, https://science.nasa.gov/resource/pioneer-venus-orbiter-map-of-venus/.

11. "What Is ALH 84001?," Lunar and Planetary Institute, https://www.lpi.usra.edu/lpi/meteorites/The_Meteorite.shtml.

12. Clara Moskowitz, "Looking for Life on Mars: Viking Experiment Team Member Reflects on Divisive Findings," *Scientific American*, April 2, 2019, https://www.scientificamerican.com/article/looking-for-life-on-mars-viking-experiment-team-member-reflects-on-divisive-findings/.

13. Anashe Bandari, "No Phosphine on Venus, According to SOFIA," SOFIA, November 29, 2022, https://blogs.nasa.gov/sofia/2022/11/29/no-phosphine-on-venus-according-to-sofia/.

CHAPTER 6

1. Dana Bolles, "Saturn," edited by SMD Content Editors, NASA, accessed January 3, 2023, https://science.nasa.gov/saturn/.

2. For a brief historical timeline on the discovery of Saturn's rings with some great sketches from Galileo, see Albert Van Helden, ed., "Saturn," The Galileo Project, http://galileo.rice.edu/sci/observations/saturn.html.

3. Dana Bolles, "Cassini-Huygens," edited by SMD Content Editors, NASA, accessed January 3, 2023, https://science.nasa.gov/mission/cassini/.

4. The study of this phenomenon is called *spectroscopy*. To play with a simulation of this process, go to https://foothillastrosims.github.io/Spectrum-Constructor/.

5. The full Cassini Mission Report is rather lengthy, but for the specific subsection on the rings, see http://tinyurl.com/4j5ynp89.

6. Dana Bolles, "Voyager 2," edited by SMD Content Editors, NASA, accessed January 3, 2023, https://science.nasa.gov/mission/voyager/voyager-2/.

7. See endnote 3 above, but for an additional resource on Cassini's final orbits, see https://science.nasa.gov/mission/cassini/grand-finale/grand-finale-orbit-guide/.

8. "Radio Occultation: Unraveling Saturn's Rings," NASA, May 23, 2005, https://photojournal.jpl.nasa.gov/catalog/PIA07873.

9. David R. Williams, "Saturnian Rings Fact Sheet," NASA, April 19, 2022, https://nssdc.gsfc.nasa.gov/planetary/factsheet/satringfact.html.

## CHAPTER 7

1. Data was an android character on the science fiction television series *Star Trek: The Next Generation*.

2. Dana Bolles, "Titan," edited by SMD Content Editors, NASA, accessed January 3, 2023, https://science.nasa.gov/saturn/moons/titan/.

3. "A Last Look at Titan," NASA, September 15, 2017, https://photojournal.jpl.nasa.gov/catalog/PIA21890.

4. For a false-color map showing all of the methane seas that have been mapped in Titan's northern hemisphere by the Cassini mission, see "Titan's North," NASA, 2013, https://photojournal.jpl.nasa.gov/catalog/PIA17655.

5. Maybe this daydream should have featured some scuba diving?

6. The International Astronomical Union has a database of all Titan geological features, in partnership with the United States Geological Survey (USGS). Look for *mare* and *lacus* at https://planetarynames.wr.usgs.gov/Page/TITAN/target.

7. "Mystery Feature Evolves in Titan's Ligeia Mare," NASA, March 2, 2016, https://photojournal.jpl.nasa.gov/catalog/PIA20021.

8. "Bright Feature Appears in Titan's Kraken Mare," NASA, November 10, 2014, https://photojournal.jpl.nasa.gov/catalog/PIA19047.

9. "Clouds in the Distance," NASA, June 28, 2005, https://photojournal.jpl.nasa.gov/catalog/PIA06241.

## CHAPTER 8

1. Rebecca Sohn, "Venus's Atmosphere: Facts about the Atmosphere of Earth's 'Twin Sister,'" Space.com, October 18, 2018, https://www.space.com/18527-venus-atmosphere.html.

2. As the MESSENGER spacecraft approached Venus, it snapped an image showing the global cloud coverage, which can be seen at https://photojournal.jpl.nasa.gov/catalog/PIA10124.

3. For live satellite images from the NOAA's Geostationary Operational Environment Satellite (GOES), see https://www.star.nesdis.noaa.gov/goes/index.php.

4. For a treatment of the Magellan data by the USGS that highlights the two continents and the surrounding lowlands, see the Venus Magellan Global C3-MDIR

colorized topographic mosaic, March 14, 2014, https://astrogeology.usgs.gov /search/map/Venus/Magellan/RadarProperties/Colorized/Venus_Magellan_C3 -MDIR_ClrTopo_Global_Mosaic_6600m.

5. "Venus from Mariner 10," NASA, June 8, 2020, https://photojournal.jpl.nasa .gov/catalog/PIA23791.

### CHAPTER 9

1. The Public Domain Review has an incredible collection of digital copies of artwork in the public domain that depict comets. The collection is called "Flowers of the Sky," published September 24, 2014, https://publicdomainreview.org /collection/flowers-of-the-sky/.

2. . Dana Bolles, "Kuiper Belt," edited by SMD Content Editors, NASA, accessed January 3, 2023, https://science.nasa.gov/solar-system/kuiper-belt/.

3. Fun fact: we have never actually observed an Oort Cloud object; we only infer that the Oort Cloud exists due to the behaviors of long-period comets. Dana Bolles, "Oort Cloud," edited by SMD Content Editors, NASA, accessed January 3, 2023, https://science.nasa.gov/solar-system/oort-cloud/.

4. Kristen Erickson, ed., "What Is a Comet?," Space Place, August 24, 2023, https://spaceplace.nasa.gov/comets/en/.

5. Most recently, Comet 17P/Holmes achieved this feat. Robert Roy Britt, "Incredible Comet Bigger Than the Sun," Space.com, November 15, 2007, https://www.space.com/4643-incredible-comet-bigger-sun.html.

6. Timeanddate.com has a table of the next Earth perihelion and aphelion dates and distances from the Sun. See https://www.timeanddate.com/astronomy /perihelion-aphelion-solstice.html.

7. Timeanddate.com has a good summary of why we have seasons on Earth, including the common misconception about distance. https://www.timeanddate .com/astronomy/seasons-causes.html.

8. "Fastest Speed Achieved by Humans," Guinness World Records, https://www .guinnessworldrecords.com/world-records/66125-fastest-speed-achieved-by -humans.

9. The Rosetta orbiter and Philae lander were incredibly successful at investigating a comet up close. The ESA mission page is https://www.esa.int/Science _Exploration/Space_Science/Rosetta.

10. A recent example of this was Comet ISON, a relativity bright comet from 2013 that disintegrated during its perihelion pass around the Sun. For more information, see https://www.esa.int/ESA_Multimedia/Images/2014/12/Comet_ISON _disintegrates.

## CHAPTER 10

1. Dana Bolles, "Asteroids," edited by SMD Content Editors, NASA, accessed January 3, 2023, https://science.nasa.gov/solar-system/asteroids/.

2. NASA's Dawn Mission was sent to the asteroid belt to study both Vesta and Ceres. The latter was formerly the largest asteroid in the asteroid belt, but was reclassified to dwarf planet status in 2006, at the same time as Pluto. Dana Bolles, "Dawn," edited by SMD Content Editors, NASA, accessed January 3, 2023, https://science.nasa.gov/mission/dawn/.

3. Dana Bolles, "Bennu," edited by SMD Content Editors, NASA, Accessed January 3, 2023, https://science.nasa.gov/solar-system/asteroids/101955-bennu/.

4. The exact size an object needs to be in order to crush itself into a sphere depends on the materials it's made of, but the largest asteroid, Vesta, is 525 km wide and is not a sphere. This is actually one of the requirements that delineates planet from dwarf planet status. https://www.iau.org/public/themes/pluto/.

5. Lonnie Shekhtman, "Surprise—Again! Asteroid Bennu Reveals Its Surface Is like a Plastic Ball Pit," NASA, July 7, 2022, https://www.nasa.gov/missions/surprise-again-asteroid-bennu-reveals-its-surface-is-like-a-plastic-ball-pit/.

6. The Understanding Global Change website by University of California, Berkeley, has a fantastic list of these processes: https://ugc.berkeley.edu/earth-systems/how-the-earth-system-works/.

## CHAPTER 11

1. Jovian translates to "of or relating to Jupiter," from Latin. In effect, it means Jupiter-like. We use it both to refer to the gas giants collectively, but also to refer specifically to Jupiter.

2. David R. Williams, ed., "Jupiter Fact Sheet," NASA, May 22, 2023, https://nssdc.gsfc.nasa.gov/planetary/factsheet/jupiterfact.html.

3. Dana Bolles, "Jupiter," rdited by SMD Content Editors, NASA, accessed January 3, 2023, https://science.nasa.gov/jupiter/.

4. The calculation is not that hard to do in an ideal situation. The equation is $v = \sqrt{\frac{GM}{R}}$, where $M$ is the mass of the planet, and $R$ is the distance from the center. For a mini tutorial on how to use it, see https://www.physicsclassroom.com/class/circles/Lesson-4/Mathematics-of-Satellite-Motion.

5. The Galileo entry probe was carried to Jupiter on the Galileo mission, which we talked about in chapter 3. Its mission was to enter the atmosphere of Jupiter and learn as much as possible.

6. The Huygens probe was carried to Titan by the Cassini mission, which we touched on twice already in chapters 6 and 7. Its mission was to land on Titan.

7. For a good primer on the electromagnetic spectrum, see https://openstax.org /books/astronomy-2e/pages/5-2-the-electromagnetic-spectrum.

8. Kristen Erickson, ed., "Why Is the Sky Blue?," Space Place, August 24, 2022, https://spaceplace.nasa.gov/blue-sky/en/.

9. Atmospheres are complicated places, but this was a short introduction to why we have layers, both here and at Jupiter. For a bit more information on the layers of all the Jovian atmospheres, see https://openstax.org/books/astronomy-2e /pages/11-3-atmospheres-of-the-giant-planets.

## CHAPTER 12

1. David R. Williams, ed., "Mars Fact Sheet," NASA, May 22, 2023, https:// nssdc.gsfc.nasa.gov/planetary/factsheet/marsfact.html.

2. While $CO_2$ does make up 95 percent of the gases in the Martian atmosphere, remember that the atmosphere is less than 2 percent as dense as Earth's. There's not that much $CO_2$ in total.

3. One of the Curiosity rover's more recent selfies has a very good view of the sky showing its color: https://mars.nasa.gov/resources/25757/curiositys-selfie-at -mont-mercou/.

4. Many different landers and rovers have caught the Martian blue sunset on camera. For a great list, see https://science.nasa.gov/solar-system/planets/mars /what-does-a-sunrise-sunset-look-like-on-mars/.

5. A presentation by Alex Innanen for the Allan I. Carswell Astronomical Observatory that covers all manner of clouds (and water) on Mars, including interesting details about the Aphelion Cloud Belt, can be seen at https://www.youtube.com /watch?v=29W4BCSVj9k.

6. It's highly recommended the reader look into sundogs! For a good place to get started, see https://earthobservatory.nasa.gov/blogs/earthmatters/2014/02/03 /reader-pics-sundogs/.

7. "NASA Explores a Winter Wonderland on Mars," NASA mars exploration, December 22, 2022, https://mars.nasa.gov/news/9326/nasa-explores-a-winter -wonderland-on-mars/.

8. Don't worry, we've thought about this: https://voyager.jpl.nasa.gov/golden -record/.

## CHAPTER 13

1. Saturn has 146 natural satellites, as of June 2023. For the current list, see https:// nssdc.gsfc.nasa.gov/planetary/factsheet/saturniansatfact.html.

2. Dana Bolles, "Hyperion," edited by SMD Content Editors, NASA, accessed January 3, 2023, https://science.nasa.gov/saturn/moons/hyperion/.

3. If you remember from chapter 10, Vesta is the largest nonspherical object in the solar system, whereas Hyperion is the largest nonspherical moon.

4. You can download the 3D model at https://science.nasa.gov/resource/hyperion -3d-model/.

## CHAPTER 14

1. The Celsius scale was created to have water freeze solid at 0°C and boil at 100°C. There are a variety of other scales with which to measure temperature, some you may be familiar with, and others not so much: https://cryo.gsfc.nasa.gov /introduction/temp_scales.html.

2. Dana Bolles, "Titan," edited by SMD Content Editors, NASA, accessed January 3, 2023, https://science.nasa.gov/saturn/moons/titan/.

3. The actual definition is a bit more complicated/precise: https://glossary.amet soc.org/wiki/Scale_height.

4. These hydrocarbon sand dunes have been observed by Cassini using radar to penetrate the clouds. They have a strong similarity to dunes we see on Earth. See https://science.nasa.gov/resource/titans-sand-dunes/.

## CHAPTER 15

1. Dana Bolles, "New Horizons," edited by SMD Content Editors, NASA, accessed January 3, 2023, https://science.nasa.gov/mission/new-horizons/.

2. Dana Bolles, "Arrokoth (2014 MU69)," edited by SMD Content Editors, NASA, accessed January 3, 2024, https://science.nasa.gov/solar-system/kuiper -belt/arrokoth-2014-mu69/.

3. For a website that has a live tracker of where New Horizons is, and keeps you up-to-date on any new science, see https://pluto.jhuapl.edu/.

4. Here's part of the story: https://www.iau.org/public/themes/pluto/.

5. However, there are probably hundreds or thousands of undiscovered objects in our solar system that meet the dwarf planet requirements. For a website with a great summary of many of the things we talked about in this book, and that provides a discussion on what a dwarf planet is, see https://science.nasa.gov/solar-system /planets/.

6. An exoplanet is a planet that is not part of our solar system. How do these fit into our current definition of planet? https://exoplanets.nasa.gov/.

7. For a very brief timeline of this story, see https://science.nasa.gov/dwarf -planets/ceres/exploration/.

8. "Apparent Magnitude," https://lco.global/spacebook/distance/what-apparent-magnitude/.

9. For a brief description of the inverse square law, see https://science.nasa.gov/learn/basics-of-space-flight/chapter6-1/.

10. David R. Williams, ed., "Pluto Fact Sheet," NASA, May 22, 2023, https://nssdc.gsfc.nasa.gov/planetary/factsheet/plutofact.html.

11. We can calculate how much brighter the Sun is than Sirius, as seen from Pluto, now that we know how the magnitude system works. At its dimmest, the Sun appears to have a magnitude of −18.3, whereas Sirius has −1.46. That's a difference of almost 17 magnitudes. Since each step accounts for a 2.5 times change in brightness, that means the Sun is almost 6 billion times brighter than Sirius, even at the Pluto distance. That's one bright Sun.

12. The mountains the protagonist climbs in the story are based on a specific mountain range on the left side of the flat icy planes (aka Pluto's Heart), seen at https://photojournal.jpl.nasa.gov/catalog/PIA19947.

13. For a true-color image of Pluto, with the infamous heart in the bottom right, see https://photojournal.jpl.nasa.gov/catalog/PIA19857.

AFTERWORD

1. Bob Ross, episode 3 of season 23 of *The Joy of Painting* (1991).

# INDEX